THE COAST OF AUSTRALIA

Andrew D. Short and Colin D. Woodroffe

Australia boasts one of the longest, most diverse and pristine coastlines in the world. Ranging from the oldest rocks on the planet to those that are still evolving today, the Australian coast is a dynamic, ever-changing suite of dramatic landforms and productive ecosystems. From iconic beaches such as Bondi and the long, unbroken sands of the Coorong, to the seemingly endless curtain of the Nullarbor cliffs along the Great Australian Bight, this book takes you on an illuminating journey around Australia's coastline. Covering 36 000 kilometres of shoreline, it details the various coastal systems that exist around the continent, including beaches, dunes, estuaries, deltas, rocky coast and coral reefs.

Written by two of Australia's leading coastal scientists, Andrew Short and Colin Woodroffe, *The Coast of Australia* provides the first comprehensive account of the Australian coast, revealing why it formed and how it continues to change.

Andrew Short is Professor of Marine Science and Geosciences at the University of Sydney. He has degrees from the University of Sydney, the University of Hawai'i and Louisiana State University. He has studied the coasts of North and South America, the Arctic Ocean, Hawai'i, New Zealand, the British Isles and Europe, as well as the entire Australian coast.

Colin Woodroffe is Professor in the School of Earth and Environmental Sciences at the University of Wollongong. He has a BA, PhD and ScD from the University of Cambridge and a Masters in Applied Science from the University of New South Wales. He has studied tropical coasts in the West Indies and the Indian and Pacific Oceans, and has undertaken research on many of the remote coasts of Australia.

The COAST of AUSTRALIA

Andrew D. Short and Colin D. Woodroffe

CAMBRIDGE
UNIVERSITY PRESS

477 Williamstown Road, Port Melbourne, VIC 3207, Australia

Cambridge University Press is part of the University of Cambridge.

It furthers the University's mission by disseminating knowledge in the pursuit of education, learning and research at the highest international levels of excellence.

www.cambridge.org
Information on this title: www.cambridge.org/9780521873987

© Andrew D. Short & Colin D. Woodroffe 2009

First published 2009

Designed and typeset by Sandra Nobes, Tou-Can Design
Cartography by Peter Johnson

A catalogue record for this publication is available from the British Library

National Library of Australia Cataloguing in Publication data
Short, Andrew D.
The coasts of Australia / Andrew D. Short, Colin D. Woodroffe.
9780521873987 (hbk.)
9780521696173 (pbk.)
Includes index.
Bibliography.
Coasts—Australia.
Coastal ecology —Australia.
Woodroffe, Colin D.
551.4570994

ISBN 978-0-521-87398-7 Hardback
ISBN 978-0-521-69617-3 Paperback

Reproduction and communication for educational purposes

Reproduction and communication for other purposes

To Professor Bruce G. Thom
for his leadership in the study and
management of the Australian coast

CONTENTS

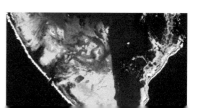

ACKNOWLEDGEMENTS

The coast of Australia continues to be an unending source of inspiration to us both. This book is based on field research we have been privileged to undertake around the entire margin of the Australian mainland and on many offshore islands. We would particularly like to acknowledge the foundation provided by the pioneering research of Professor J. L. (Jack) Davies, who first described the geographical variation of coasts around Australia. It has been our good fortune to work closely with Professor Bruce Thom, who has led the way in understanding recent sea-level history and evolution of Australia's sedimentary shorelines, and has played a unique role in the development of policies that relate to the coast, incorporating geomorphological principles into coastal management. It is our pleasure to dedicate this volume to Bruce.

The ideas we explore in this book have been shaped by the insights and assistance provided by our many colleagues and students over the years. The outstanding contributions of John Chappell, Jim Coleman, David Hopley, Roger McLean, Peter Roy, Choule Sonu, David Stoddart and Don Wright have been particularly influential. Their contributions continue to provide the directions for ongoing and future research. We have been privileged to work with, and appreciate the assistance, insights and companionship of, Rob Brander, Brendan Brooke, Ted Bryant, Peter Cowell, Mark Dickson, Ian Eliot, Peter Harris, Nick Harvey, Andrew Heap, Patrick Hesp, Brian Jones, David Kennedy, Graham Lloyd, Gerd Masselink, Colin Murray-Wallace, Kevin Parnell, Ava Simms, Scott Smithers and Ian Turner, as well as those of the many other friends, colleagues and students with whom we have worked. We also thank Brett Williamson, Greg Nance and all the crew at Surf Life Saving Australia who supported the Australia-wide beach surveys.

We are indebted to Peter Johnson, who has supported us over the years with his graphic skills, for drawing all the artwork for this book. We thank Phil Coleman for reading and commenting on some of the early drafts, and John Marthick for assistance with imagery.

We thank the team at Cambridge University Press for the keen interest and encouragement they have shown throughout the evolution of this book; to Kim Armitage for supporting the idea, Pauline de Laveaux for overseeing its development, Renée Otmar for editing and improving the flow of the book, Sandra Nobes for the design and to Jodie Howell for project management. This team has turned our draft into this publication.

The maturing of our studies of the coasts and the writing of this book could not have occurred without the unstinting support of our families. We particularly thank our wives, Julia Short and Salwa Woodroffe, for their support.

PROLOGUE

Australia has one of the longest, most ancient, diverse, pristine and spectacular coastlines in the world. The shoreline ranges in age from the oldest rocks on the planet, to those that are still being formed today through the addition of sediment from rivers or the breakdown and accumulation of organisms such as coral or shell. The continent experiences a wide range of climates around its coast, from tropical monsoon in the north through to temperate–humid and arid climates in the south and west. Half of the mainland's more than 30 000-kilometre coast consists of over 10 000 sandy beaches; the remainder is predominantly rock, much of which is sculpted into impassably steep cliffs. Australia's coast is home to many plants and animals that are found nowhere else on Earth. Across the northern shoreline are the most extensive coral reef systems in the world, along with some of the largest areas of undisturbed mangrove forests and extensive tropical seagrass meadows. The southern half of the continent is exposed to the world's most energetic wave climate and supports the world's most extensive temperate seagrass meadows and shelf carbonate system, which has supplied massive volumes of marine sand to build the beaches and dunes.

The Australian coastline is a dynamic, ever-changing suite of dramatic landforms

and productive ecosystems. Coastal water-bodies include more than 1000 estuary and delta systems where rivers reach the coast, along with over 1500 smaller streams and tidal creeks. The majestic Port Jackson, which includes Sydney Harbour, is one of the many estuaries located along the east coast. By contrast, on the west coast lies Shark Bay, a large, arid, salty bay that contains rare colonies of algal stromatolites, the modern equivalents of one of the oldest life forms on earth – these are our living 'fossils'. Across northern Australia there are many estuaries with broad tidal flats covered with mangroves. Here, the tidal range can reach several metres, with the world's third-highest tides found in King Sound, Western Australia. The southern coast is pummelled by large waves, and the mouth of our largest river, the mighty Murray, is shaped by wave action, which on occasion seals off the much-reduced river flow.

Australia's beaches include the iconic Bondi Beach in Sydney, long unbroken beaches such as the Coorong in South Australia, Ninety Mile Beach, which flanks the Gippsland lakes in Victoria, and Eighty Mile Beach, located near Broome in Western Australia. The strong westerly winds in the south of the continent and the southeast trade winds that affect the north coast have built some of the largest and most extensive

coastal dune systems in the world, including the world's largest sand island, Fraser Island.

Vast stretches of coastline are composed of cliffs. The seemingly endless curtain of the Nullarbor cliffs along the Great Australian Bight, and the scenic, indented cliffscape near Port Campbell contrast with the stately sandstone cliffs that flank the heads and line the shores of Sydney, the dolerite columns of the southern shore of Tasmania and the rugged granite cliffs forming numerous capes and headlands in between. Bold, rocky coasts in the tropics support fringing coral reefs. The Great Barrier Reef is the most extensive coral reef in the world, containing a multitude of rugged volcanic islands and secluded coral cays across its broad lagoonal shelf. The pristine Ningaloo reef fringes the shore along the arid North West Cape, and more remote and lesser-known reefs occur along the north coast and offshore.

In this book we take you on a journey to admire these magnificent landforms and to understand why they formed and how they change, continually responding to waves, tides, wind and other climatic and oceanographic factors. Many of our coasts bear the imprint of humans. On some parts of the coast, traces remain of use by the earliest inhabitants of the Australian continent. In other places, high-rise apartments support high densities of inhabitants, and the desire for a 'seachange' lifestyle sees many more people migrating to the coasts each year. Other coasts remain almost untouched, even today, although the indirect effects of climate change are threatening them and seem certain to be felt in coming years. Most Australians have seen little of, and know little about, these national treasures. Come with us on a journey of discovery and share our wonder for the magnificent coast of Australia.

EVOLUTION OF THE AUSTRALIAN COAST

| Introduction

THIS BOOK PROVIDES a comprehensive overview of the more than 30 000-kilometre-long Australian coast (see Figure 1.1). It is based on the latest scientific investigations and thinking, and aims to leave you, the reader, with a clear understanding of the coastline, including its geological background and evolution, and the processes and ecosystems that operate around the coast. A major part of the book is devoted to describing each of the systems that make up the coast, starting with the rivers, estuaries and deltas, and followed by the thousands of beach systems and the dune systems that back most of the beaches. We describe the extensive rocky shores that separate many of the beaches, then the coral reef systems that fringe much of northern Australia. We conclude with an overview of human impact on the coast, including the potential impacts of climate change.

Defining the Australian coast

The length of the Australian coastline depends, first, upon how we define coastline. In this book, when we use the term 'coastline', we mean the broader coastal zone. We may also refer to the 'shoreline', by which we mean the point at which the ocean meets the land. The length of the coastline – or shoreline – also depends on the precision with which we measure it. For example, if we were to divide the coastline into segments, each 1 kilometre long, the length of the coastline around the mainland and Tasmania would

State/Territory	Mainland area (km²)	Island area (km²)	Total area (km²)	Mainland length (km)	Island length (km)	Total length (km)
Western Australia	2 526 786	3089	2 529 875	12 889	7892	20 781
Queensland	1 723 936	6712	1 730 648	6973	6374	13 347
Northern Territory	1 335 742	13 387	1 349 129	5437	5516	10 953
South Australia	978 810	4672	983 482	3816	1251	5067
Tasmania	64 519	3882	68 401	2833	2049	4882
Victoria	227 010	406	227 416	1868	644	2512
New South Wales	800 628	14	800 642	2007	130	2137
ACT*	2430	1	2431	54	3	57
Australia	7 659 861	32 163	7 692 024	35 877	23 859	59 736

Table 1.1 Land area and length of coastline for Australia, by state/ territory, as determined from digitisation of the 1:100 000 topographic maps

* Australian Capital Territory: coastline relates to Jervis Bay Territory
Source: Geoscience Australia, www.ga.gov.au.

measure 30 270 kilometres. However, if we were to add Australia's islands of more than 12 hectares, the coastline would measure 47 070 kilometres.

If we were to measure the shoreline, which includes mangroves but excludes coral reefs, using 1:100 000 topographic maps (see Table 1.1), the length of the Australian shoreline would be 35 877 kilometres for the mainland alone, or 59 736 kilometres including Australia's islands. These estimates would increase if the shoreline were determined from larger-scale maps, and if we included the intricate perimeter of coastal waterways. They would also increase if the shorelines of Australia's island territories, including the Australian Antarctic Territory, were included.

As well as possessing one of the longest national coasts in the world, Australia claims the third-largest Exclusive Economic Zone (EEZ) in the world, with an area of 8.1 million square kilometres. This is equivalent to about 2.2 per cent of the world's ocean area and larger than Australia's land area (see Table 1.1).

The EEZ adjacent to the Australian mainland covers more than 6 million square kilometres, and is augmented by the extensive EEZ around its small island territories such as the Cocos (Keeling) Islands, Christmas Island and Heard and McDonald Islands.

Structure of this book

The remainder of this book traces the nature of the Australian shoreline, its beaches, dunes and rocky coast, the intricate river mouth and estuary systems and the myriad coral reefs. In this chapter we outline the geological evolution of the Australian continent – the most ancient on the planet. We describe the continent when Australia lay 3000 kilometres further south, attached to the supercontinent known as Gondwanaland. The Australian landmass as we know it became recognisable after it detached from Antarctica and began its slow and continuous drift to the north. We conclude the chapter by describing the impact of global cooling, the beginning of the ice ages and associated major changes in sea level that helped to form the present coast and shoreline.

In Chapter 2 we look at the contemporary processes that form and rework the coast, beginning with the Australian climate, which delivers the rainfall needed to maintain our rivers and to distribute sediment to parts of the coast. Across the oceans, cyclones and strong winds generate wave climates that provide the energy to build beaches and erode rocky cliffs. The ocean also delivers the tides that range from very small on the south coast to some of the world's largest in the north. The oceanic and coastal wind systems also drive the major ocean currents and build the sand dunes along the coast. In total, these processes control the evolution of the coast.

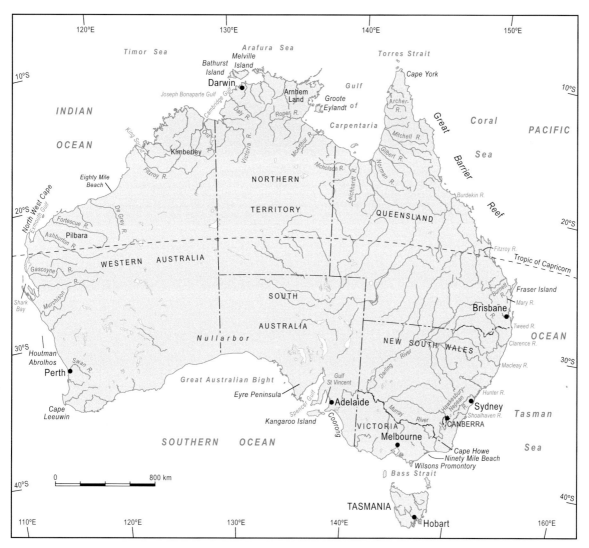

Figure 1.1 *Map of Australia, showing some of the major locations mentioned in this book*

Figure 1.2 *Arndani lagoon, in eastern Arnhem Land, exemplifies the diversity of the Australian coast. It is bordered on its seaward side by seagrass meadows, with beaches, dunes and low rocky sections forming the shoreline. An inlet and tidal channel connects the lagoon to the Gulf of Carpentaria, while the lagoonal shoreline has a fringe of green mangroves, backed by the white salt flats.* Photo: A.D. Short

In Chapter 3 we describe the variety of coastal ecosystems that are such a vital part of the coast. First, we examine the land-based, coastal dune vegetation and coastal freshwater wetlands, then the intertidal samphire (coastal herbs), salt marshes and mangroves, and the subtidal seagrasses and coral reefs. We conclude by looking at the rich ecology of the sandy beaches and rocky shore.

In the following five chapters we examine each of the major coastal systems, which together make up the Australian coast (see Figure 1.2).

In Chapter 4 we discuss the estuaries and deltas associated with river mouths, streams and tidal creeks, and their distribution, habitats and evolution. Following this we take you on a tour of the coast, highlighting some of the major estuarine systems and their regional variability.

What makes a beach? In Chapter 5 we look at Australia's more than 10 000 beach systems that occupy half of the open coast. We describe the 15 different types of beaches around Australia, their nature and distribution. Australia is renowned for its beaches, and in this chapter we finish with an overview of beaches from around Australia, exploring why Australia has the world's best beaches. In Chapter 6 we describe the coastal sand dunes that back many of the beaches and the associated larger sand barrier systems, which fringe 40 per cent of the coast.

While approximately half the Australian coast consists of soft sandy shorelines – the beaches – much of the other half is composed of hard, resilient rock. We therefore begin Chapter 7 with a discussion of the types of rocks and their distribution around the coast, and then explore the dynamic interaction between waves, tides and rocks that leads to the erosion and evolution of our rocky coast. In this chapter we again conclude with a tour of the coast, highlighting some of the more spectacular sections of Australia's rocky coastline.

Australia's world-famous coral reef system is our focus in Chapter 8, which begins with a description of reef structure and the processes that control reef evolution. We then examine the types of reefs that occur around the coast, before looking at the distribution of reefs around northern Australia and some of its major offshore islands.

In the final chapter, Chapter 9, we reflect on the influence and impact of humans on the Australian coast. We start by describing the arrival of Indigenous Australians and, later, European explorers and the initial coastal settlements, which were all located in estuaries. We assess the increasing impact of humans on the coast, the way in which the coast is managed and some present-day threats to the coast. We conclude with a discussion of the coastal impact of climate change and the future prospects for the Australian coast.

Geological evolution

The Australian coast surrounds an ancient continent, and yet the coastline is also continually being renewed. In some places the coast is composed of some of the oldest rocks on Earth, while in others it is built from sediments that are reworked every day. The type of coast that forms is intimately related to the geology of the continent. In order to understand the nature of the present-day coast, we need to consider the evolution of the Australian continent, its northwards drift and fluctuating climates, and the changes in sea level it has experienced.

Understanding the 'flat' continent

The Australian continent has been exposed to weathering for so long, and with no recent mountain building, that it has eroded to become the world's oldest, lowest and flattest continent, with an average height of just 330 metres. Its latitude, centred at 30° South under the dry, subtropical high-pressure system, also makes it the driest inhabited continent,

Figure 1.3 *The heavily jointed and deeply dissected coast at Cone Bay in the western Kimberley is part of the ancient Australian Craton. The rocks are composed of 1800-million-year-old Proterozoic sandstones.* Photo: A.D. Short

Figure 1.4 *Australia is made up of the ancient Pilbara–Yilgarn cratons, which merged by 2200 million years ago, the uplifted Kimberley Basin and the Arunta–Gawler cratons, which were added by 1800 million years ago to form the Precambrian Australian craton. Successive episodes of mountain building associated with the Tasman Fold Belt (400 million years old) and later the New England Fold Belt (200 million years old) completed the geological evolution of the Australian continent, at that time still part of Gondwanaland.*

with an average rainfall of 465 millimetres, resulting in extensive arid regions both in the interior and along parts of the south and west coast.

The western core of the continent is composed of some of the oldest surviving rocks on the planet, 3500 million years in age, which form the vast Pilbara and Yilgarn cratons (ancient portions of the Earth's crust). These rocks form outcrops along much of the irregular Pilbara coast, and along the south and southwest Western Australian coast, where massive, sloping granite headlands dominate the shore. To the east, the continent is composed of rocks formed during periods of continental welding and mountain building over the past 2000 million years. In the south, the Gawler Craton, in central South Australia, formed and attached by 2000 million years ago, and the Northern Australian or Arunta Craton attached to the western cratons more than 1800 million years ago. These four giant cratons, the Pilbara, Yilgarn, Northern/Arunta and Gawler cratons, together with the uplifted Kimberley Basin (Figure 1.3), combined to form the Australian Craton, which occupies all of western and central Australia, including much of Cape York Peninsula (see Figure 1.4). Between 1800 and 650 million years ago, the eastern boundary of this craton buckled to form the Adelaide geosyncline (a major downwarp in the Earth's crust, in which great thicknesses of sedimentary sequences have accumulated), with the eastern side of the geosyncline, known as the Tasman Line, becoming the eastern boundary of the evolving 'Australia' (see Figure 1.4).

The formation of the third of Australia east of the Tasman Line commenced in the late Cambrian period (600 million years ago) and involved a series of orogens, or episodes of mountain building. Formation of the Adelaide Geosyncline (1400–560 million years ago) was followed 500–330 million years ago by development of the Kanmantoo Fold Belt, which borders eastern South Australia and extends south through western Victoria and across Bass Strait to include the western half of Tasmania. Then a broad zone, known as the Tasman Fold Belt, evolved extending to the present east coast. This includes the granite-rich Thompson-Lachlan Fold Belt, which extends north from eastern Tasmania, through the Snowy Mountains into western New South Wales and central Queensland. It formed 500–230 million years ago, and also includes the New England Fold Belt, which extends north from the Hunter region to eastern Cape York Peninsula and formed 400–230 million years ago. Wedged between these two mountain belts is the 1800-kilometre long Sydney–Bowen Basin, into which sediments hundreds of metres thick were deposited 270–180 million years ago, including the rich basal coal units. All the time this massive eastern landmass was being added to the continent, Australia was attached to Antarctica and formed the northern part of Gondwanaland.

During the formation of the Tasman Fold Belt, buckling along the western margin of the continent

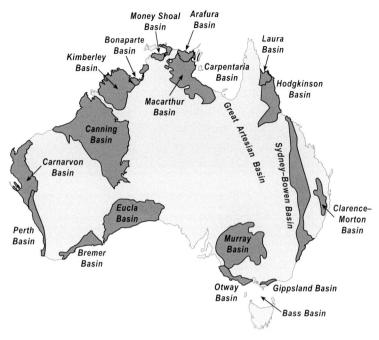

Figure 1.5 *Australia's coastal sedimentary basins occupy over half of the coastline. All were formed from buckling of the Earth's crust, and all continue to infill with sediments.*

resulted in the formation of four major sedimentary basins, which today occupy much of the west coast and continue to fill (see Figure 1.5). In the northwest is the 500-million-year-old Bonaparte Basin, which today is occupied by Cambridge Gulf. The massive Canning Basin, which extends between Port Hedland and King Sound, reaching more than 1000 kilometres inland, was also initiated 500 million years ago. The elongate Carnarvon Basin, located between the Murchison River and Dampier Archipelago, has been filling with sediments for the past 400 million years. The narrow, 1000-kilometre-long Perth Basin extends south from the Murchison River to Augusta; it commenced filling 450 million years ago.

Separating from Gondwanaland

The breaking up of Gondwanaland commenced in the northwest 180 million years ago, when India and Africa separated from Antarctica and Western Australia, opening up the Indian Ocean and forming the west Australian coast (see Figure 1.6). This rifting (gradual separation of two continental or oceanic plates) continued counter-clockwise around Australia, forming first the west coast, then the southern coast as Australia itself separated from Antarctica 120 million years ago and began its slow northward migration. This was followed by rifting in the southern Tasman Sea 85–60 million years ago, which separated New Zealand and the Chatham Rise from southeast Australia. As a result, the Tasman Sea opened and formed the coast along southeastern Queensland–New South Wales, eastern Victoria and eastern Tasmania. The rifting continued northward, opening the Coral Sea 55–50 million years ago and forming the rest of the Queensland coast. As the continent moved north, buckling, mountain building and volcanic activity on the northernmost margin formed the New Guinea highlands, which was also part of the Australian plate – uplift continues there to this day.

While the continent was rifting and moving it caused regional buckling around its perimeter (see Figure 1.5), which formed a series of shallow coastal basins, including the large Carpentaria Basin (180–100 million years old) in the north. Across the south coast the Eucla (Nullarbor) (160–0 million years old), Otway–Murray (170–40 million years old), Otway (50–20 million years old), Gippsland (35–2 million years old) and Bass basins (120–0 million years old) also formed. Each of the basins flooded, gradually filling with sediments and then subsequently uplifted, though parts of the Gippsland and the entire Bass Basin have been submerged by the recent postglacial sea-level rise, described later in this chapter.

Along the east coast, during the opening of the Tasman and Coral seas, there was subduction of the oceanic plate under the eastern seaboard's continental plate, which resulted in the gradual uplift of the eastern highlands between 85 and 60 million years ago. The net result was the formation of the Great Dividing Range, which extends for 3000 kilometres down the east coast, from Cape York to central Victoria. This event finalised the general outline of the Australian coast. It was now separated from Antarctica and the Tasman Rise (including New Zealand), surrounded by the southwest Pacific, Southern and Indian oceans, and drifting north at about 7.3 centimetres per year.

Australia's northward drift has had two additional major impacts on the continent and coast. First, beginning about 40 million years ago, it increasingly placed Australia under the influence of the great subtropical high-pressure systems, the source of the continent's aridity. Second, it opened up the Southern Ocean, enabling the strong westerly

AUSTRALIA'S MOST RECENT VOLCANO

The most recent volcanic activity in Australia occurred at Tower Hill in western Victoria. The 'hill', an extinct volcano, last erupted 25 000 years ago, building the ash cone that can be seen today, and sending a river of lava several kilometres south to the coast. The lava forms the low basalt points, rocks and reefs between Killarney Beach and Port Fairy, the youngest igneous (volcanic) rocks in Australia. Tower Hill was declared Victoria's first national park in 1892. It is now a State Game Reserve.

winds associated with the prevailing mid-latitude cyclones to blow across the longest fetch of ocean in the world. These westerlies were named 'the roaring forties' and 'raging fifties', after the latitudes at which seafarers encountered them. High waves and strong winds have battered the southern half of the continent ever since.

Climate change and sea-level fluctuations

Climatic changes and the associated adjustments of sea level that have occurred over the past few million years have exerted a major influence on how the Australian coast has evolved. Gradual global cooling commenced about 10 million years ago, primarily as a result of the opening of the Southern Ocean. Falls in global temperature triggered the accumulation of ice caps on Antarctica and Greenland. By 2 million years ago, both the cooling and accumulation of ice caps was sufficient for ice sheets to begin to form on Eurasia (centred on the Baltic Sea) and North America (centred on Hudson Bay). These cooler periods are termed 'glaciations'. They result in a lowering of sea level (up to 120 metres below the present sea level) because of the amount of water 'locked up' in the glacial ice sheets – this is known as a sea-level 'lowstand'. The warmer periods of these ice-age cycles are called 'interglacials'. As in the present interglacial, which is usually called the 'postglacial', the Eurasian and North American ice sheets melt completely, but the Greenland and Antarctic ice sheets do not. As a result, the sea level rises to what is called a sea-level 'highstand' – at or close to our present sea level.

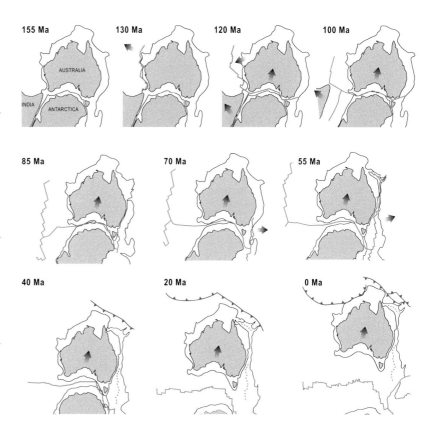

Figure 1.6 *The separation of Australia from the supercontinent Gondwanaland commenced in the northwest 155 million years ago, forming the Western Australian coast by 130 million years ago. The Indian subcontinent began separating around 120 million years ago. By 100 million years ago, both India and Australia were moving north (arrows) and Australia's southern coast was formed. By 40 million years ago the Australian continent began colliding with, and subducting beneath, the Pacific plate north of New Guinea. Australia has continued to move north at about 7 centimetres per year, opening up the Southern Ocean and colliding with the Indonesian Archipelago and oceanic plate north of New Guinea (red line). Blue lines indicate active submarine spreading; dashed blue lines inactive spreading; and black outlines the edge of the continental shelves and plates.*

Cycles of ice-sheet growth and decay

The past 2.5 million years represent the Quaternary Period, which consists of the Pleistocene and Holocene epochs (see box 'The Cenozoic era'). It is a period dominated by the ice ages, during which time the Eurasian and North American ice sheets accumulated during glaciations, followed by melting of the ice sheets and sea-level rise to interglacial highstands. The growth of these ice sheets has been gradual, but the decay is quicker, taking in the order of 10 000 years. The cause and timing of these cycles are linked to regular variations in the Earth's orbit: variation in its eccentricity (the shape of the Earth's elliptical orbit around the sun) at periods of 100 000 years; its obliquity (the angle of tilt of the Earth's axis) with a cycle of 41 000 years; and the precession of the equinoxes (the wobble, or rotation, of the orbital plane) that recurs every 22 000 years. These astronomical variations have effects on the balance of heat at different latitudes and, hence, global temperature. They result in the growth and decay of the Eurasian and North American ice sheets in a cyclical manner that reflects these 20 000, 40 000 and 100 000-year periodicities.

The record of successive interglacials and associated sea-level highstands is particularly well preserved as a series of coastal sand barriers located on the high-energy southeast coast of South Australia. The barriers parallel the present coast for over 200 kilometres and extend one after the other up to 300 kilometres inland. The age of each of these barriers – determined by luminescence dating, a technique that measures trapped electrons by releasing their energy in the form of light – can be correlated with successive sea-level highstands (see Figure 1.7). The Last Interglacial shoreline can also be clearly observed at several locations along the coast of Western Australia, where it remains as a raised coral reef. The best sequences of raised coral reefs, providing the most useful records of past sea-level changes, occur on shorelines that have been tectonically uplifted by earthquakes, as around the Indonesian island arc, on which there is a staircase-like series of raised reefs or terraces, which increase in height with age. A sea-level reconstruction has been undertaken based on the ages of such reefs on the Huon Peninsula in Papua New Guinea (determined by uranium-series dating). Dating of terraces from several sites at which there have been different rates of uplift has enabled a history of sea-level positions to be developed, and this has been linked with the record of global temperature analysed from deep-sea cores (determined by a geochemical technique called 'oxygen isotope analysis'). Figure 1.7a shows a reconstructed, sea-level curve derived from the Huon Peninsula which indicates the probable pattern of sea-level change over the past two glacial–interglacial cycles; that is the past 240 000 years.

Patterns of sea-level fluctuation

The details of past sea-level fluctuations continue to be refined. Dating the time and depth of sea-level

THE CENOZOIC ERA

Tertiary Period	65–2.5 million years ago	period of gradual global cooling
Quaternary Period	2.5–0 million years ago	beginning of ice ages (glacials and interglacials)
Pleistocene Epoch	2.5 million years to	(approximately 20 major ice ages) 10 thousand years ago
Holocene Epoch	10 thousand years ago to present	(period of postglacial warming and present interglacial)

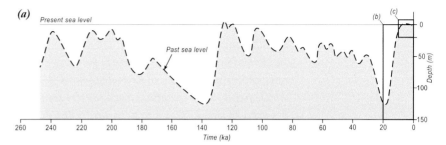

lowstands is difficult because the fossil shorelines lie drowned on the continental shelf. However, evidence from northern Australia indicates that during the Last Glacial maximum (20 000 years ago) sea level reached 120–140 metres below present. Figure 1.7b shows evidence for the pattern of sea-level rise during the postglacial period of warming derived from around the tropics. Following the lowstand, global temperature rose and ice melt commenced about 18 000 years ago. This continued rapidly for an 11 000-year period, during which time sea level rose at an average rate of around 10 millimetres per year. Two particularly rapid phases of sea-level rise, termed meltwater pulses, occurred, during which sea-level rise exceeded 20 millimetres per year. Most of the ice melt appears to have been complete by around 7000 years ago, and sea levels around Australia stabilised at a level close to present about 6500 years ago.

Isostatic adjustment

In addition to the global rise and fall in sea level, which is called 'eustatic', two other adjustments cause regional changes in the level of the sea relative to the land, called 'isostasy'. First, the accumulation of vast ice sheets of several kilometres thickness imposes a considerable load over the poles, and the Earth adjusts to the redistribution of such loads by a process known as glacial isostasy. The crust is depressed by hundreds of metres under the weight of ice; when the ice sheets melt the crust rebounds over a period of several thousand years. Areas that were loaded with an ice sheet, such as Scandinavia and northern Canada, are still experiencing uplift in response to the melting of the ice, and gradual rebound of the crust.

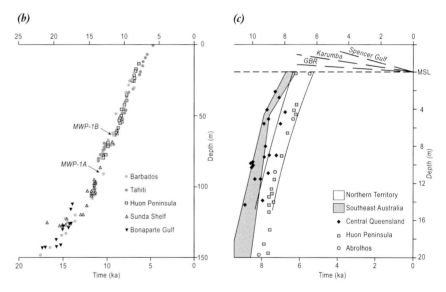

Figure 1.7 *Quaternary sea-level change in the Australian region [ka = thousands of years]: a) Sea-level change over the past 240 000 years, based on dated, raised reefs on Huon Peninsula, tuned to the deep-sea isotope record; b) the pattern of sea-level rise from the Last Glacial maximum, as reconstructed from cores from Joseph Bonaparte Gulf in northern Australia, in comparison with fossil reefs around Barbados, Tahiti and the Huon Peninsula [MWP = meltwater pulse]; c) Holocene sea-level data from around Australia showing early Holocene rise from several sites around Australia, which shows that the sea reached present level about 6500 years ago.*

The second is hydro-isostasy, which refers to the adjustment of the ocean floor due to the addition and loss of up to 150 metres of water column. This is most pronounced on the shallow continental shelves, which may be exposed during a sea-level lowstand and drowned by more than 100 metres of water during a highstand. The weight of this water results in a depression of the shelf and upward tilting of adjacent shoreline by a metre or two. Australia remained largely free of ice during the glacial periods (except for minor local ice in the Snowy Mountains of southeastern Australia and the Central Highlands of Tasmania), and hence is largely remote from isostatic adjustment to ice loads. However, the shorelines around the continent have experienced much more subtle hydro-isostatic movements, with the 6500-year shoreline along much of the northern and parts of the southern coast now uplifted by 1–2 metres above sea level.

Radiocarbon dating, which measures the decay of radioactive carbon isotopes from organic material to date deposits, provides a particularly effective means of dating fossil shorelines formed during the Holocene. Figure 1.7c shows some of the evidence for sea-level highstands around Australia resulting from hydro-isostatic adjustment across the continental shelf. One of the first detailed sea-level envelopes was derived from numerous dates on shells from cores through beach foredune ridge sequences in south-eastern Australia. This has since been supplemented by dates from muddy deposits (particularly mangrove mud) from northern Australia, and from dates on reef growth from the Abrolhos reefs off Western Australia and Queensland's Great Barrier Reef. A similar pattern of sea-level rise is also recorded from Spencer Gulf in South Australia, based on assemblages of microfossils, particularly foraminifera.

The nature of the coast that has developed since sea-level stabilisation around 6000 years ago, varies from place to place around Australia, and is associated with differences in the extent that the sea has fallen relative to the land because of hydro-isostasy. For example, Karumba, in the southeastern Gulf of Carpentaria, contains a sequence of low-energy, shelly beach ridges (ridges deposited by waves, usually on sand flats), and cheniers (ridges deposited on mud flats), which have been interpreted as evidence of nearly 2 metres of sea-level fall. There is abundant evidence from flat-topped intertidal coral colonies, called micro-atolls, that the sea was relatively higher than present along most of the mainland coast inside the Great Barrier Reef. Similar evidence for a slight relative fall of sea level over the past few thousand years has been observed on the Western Australian coast, from the Abrolhos islands and around Perth. Evidence for higher sea level has been more ambiguous along the coast of southeastern Australia, where it appears that the sea reached only slightly above present around 6000 years ago. In South Australia, a highstand of more than 4 metres above present occurred at the northern end of Spencer Gulf but was only around 2 metres at the southern end. This period of the past 6000 years, and the detail of local sea-level trend at sites around Australia, is important because that is the time during which most of the unconsolidated sandy and muddy shorelines around the coast have evolved, and sea level has exerted a particularly strong control on the landforms that have formed and are continuing to develop. It also provides an indication of how shorelines may respond if sea level rises in future, a concern that is examined in Chapter 9.

Depositional and erosional events

The modern Australian coast is therefore a complex outcome of a wide range of geological events extending back 3500 million years. These events involved both mountain building and formation of sedimentary basins, which fill with sediments eroded from the mountains, as well as volcanic activity. The present coastal margin has formed since separation from Gondwanaland, and the subsequent rifting in the Tasman and Coral Seas. Superimposed on this framework, during the drift northwards, have been a series of depositional and erosional events as the climate has changed and the sea level has fluctuated, particularly over the past few million years. The most recent events associated with the postglacial 120-metre rise in sea level have left the greatest legacy and have resulted in the location and shaping of the present Australian coast. Episodes of this geological history can be seen where the coast is rocky, but much of the coastline is built from unconsolidated sediments, and it is these we describe in this section.

During the past 2.5 million years, the Australian coast has been subject to repeated fluctuations in sea level, with at least 20 sea-level highstands close to present sea level, recurring roughly every 100 000 years. Because the Australian continent is relatively stable, sea level has returned to a similar elevation at each highstand, and valleys have flooded repeatedly to form estuaries. Rivers gradually have deposited sediment and built deltas; waves and wind have built beach and dune systems along exposed sections of the coast; rocky coasts have eroded to form sea cliffs; and in the tropics coral reefs have formed with reef tops at sea level.

When sea level falls these coastal landforms are left stranded; they may be up to 100 metres above sea level, during lowstands, tens to hundreds of kilometres inland, and far removed from coastal processes. One hundred thousand years later, when sea level again rises to a highstand, many of these dormant coastal landforms can be partly or fully reactivated by waves, tides and winds. As a consequence, past relict landforms formed during earlier sea-level highstands are located around the entire Australian coast. Some are massive and dominate the coast; many lie just inland of the modern coast. Some are blanketed by more recent coastal sediments, whereas others are eroded by the rising sea and are no longer preserved.

The most extensive Quaternary coastal sedimentary deposits in Australia occupy the lower half of the Murravian Gulf, which extends from the southeast of South Australia into southwestern New South Wales and western Victoria (see Figure 1.8). Sediments delivered by the Murray–Darling river system, combined with carbonate sediments eroded from the continental shelf and delivered by waves, are reworked at each highstand by strong southerly waves and winds to form 100-kilometre-long, 20–50-metre-high coastal sand barriers up to 5 kilometres wide. Slow uplift of the region at 7 centimetres per 1000 years (or 7 metres per 100 000 years) has helped to preserve the highstand barriers above sea level. The innermost of more than 20 barrier systems is more than 2 million years old, lies 500 kilometres inland and is up to 60 metres above sea level.

A series of up to 12 stranded barriers, between Robe and Naracoorte in South Australia, record the most recent interglacial highstands. Locally the barriers are known as 'ranges', separated by

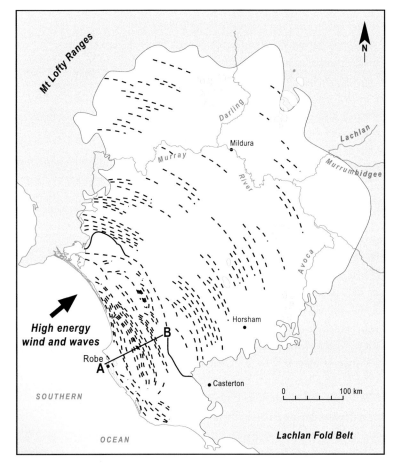

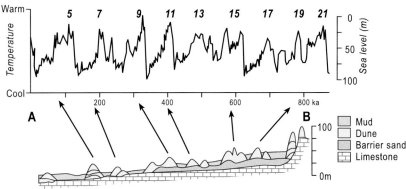

intervening interbarrier depressions composed of muddy lagoonal deposits, or 'avenues', now largely drained to form rich farmland. This is one of the longest sedimentary records of a sequence of Quaternary highstands anywhere in the world. Luminescence dating of the barrier sands indicates that they correlate with warmer interglacial conditions during which the mixed calcareous and silica sands were reworked into barriers by the high-energy waves typical of this coast. Within the ranges, beach deposits are overlain by lithified dune deposits up to 40 metres high and form barriers that may be a few kilometres wide. They have been well preserved because of the gradual uplift and weak cementation of the calcareous-rich beach and dune sediments.

Coastal sediments

Since half the Australian coast consists of unconsolidated sediments, which are potentially mobile, it is critical to understand the nature, sources and sediment dynamics of these systems, in order to manage and maintain the coast. Unconsolidated

Figure 1.8 During the past 2 million years, high-energy waves and strong southerly winds have reworked sediments of the Murray–Darling river system with shelf carbonate to gradually fill the Murravian Gulf with a 500-kilometre wide plain of Quaternary coastal sand barriers. Each barrier records a sea-level highstand. The barriers are being slowly uplifted and the inner barriers have been substantially reworked by strong westerly winds during cooler glacial periods. The outermost set forms the so-called 'ranges' and 'avenues' of southeast South Australia, as shown in the lower cross-section (A–B), with the sand barriers (ranges) up to 40 metres high in yellow, and the intervening depressions (avenues) in green.

deposits take the form of wave-deposited beaches and backing dunes deposited by winds; estuarine deltas and mud basins; and mangrove-lined tidal sand and mud flats. The composition of Australian coastal sediments is related to their origin. Most sediments are either *terrestrial*, derived from land-based rocks, which are eroded and delivered by rivers and streams to the coast, or *marine*, the remains of calcareous organisms that form hard carbonate shells or skeletons that are washed shoreward by waves and in the process are usually broken down into gravel, sand and finer particles.

Terrestrial sediments

Terrestrial sediments (land-derived, and also called 'terrigenous') are predominantly quartz sand grains, the most abundant mineral on the planet and also one of the most resilient. Quartz is an important component of granite rocks, and all quartz is ultimately derived from granite. The granite also supplies small percentages of equally resilient heavy minerals (such as rutile, ilmenite and zircon) and softer minerals such as feldspars. The softer, less stable minerals are ground down or dissolved into mud and silt, but quartz and heavy minerals are hard, resilient and remain intact as sand grains.

The composition of the sediments varies regionally depending on the geology of the river catchments. Some regions, like the hinterlands of eastern and western Australia, are abundant in quartz-rich granite and supply quartz sand grains, while the rivers draining into the Gulf of Carpentaria deliver large quantities of finer material. Across much of the arid south and central west, where there are no rivers and thus no terrigenous sediment presently reaches the coast, the sediments that occur are primarily marine in origin.

Marine sediments

Marine sediments are derived from the skeletal breakdown of calcareous organisms. They are generally referred to as 'carbonate sediments', as they are composed primarily of calcium carbonate. In Australia, there is a contrast between tropical coral reef sediments and the carbonate sediments formed on cooler, temperate shores. In the tropical north, coral reefs produce coral, algae and shell, and the broken debris from these organisms may be delivered to the adjacent island and mainland shores. Carbonate sediments comprise the principal fraction on the outer reefs of the Great Barrier Reef, but terrestrial sediments contribute most to mainland beaches, and both contribute to the continental shelf in between. In the south and west, sheltered sections of coast have productive temperate seagrass meadows growing in shallow water close to shore. These provide a habitat for a range of molluscs, foraminifera and algae that are eroded and swept, at times intact, to the shore to form shell-rich, low-energy beaches. On the most exposed southern and western coast, a range of organisms (molluscs, algae, bryozoans, foraminifera and echinoids) live on the deeper inner continental shelf, and are also detached and transported shoreward from depths as great as 70 metres. They are usually broken down to fine–medium sand, to provide most of the sand for the wave-dominated, high-energy beaches between Wilsons Promontory and North West Cape.

Figure 1.9 illustrates the distribution of beach sediments according to their source; that is,

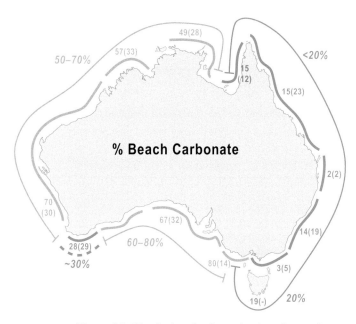

Figure 1.9 *Distribution of carbonate beach sands around Australia; inner numbers indicate regional percentage of carbonate with standard deviation in brackets. The blue shading indicates the humid coastal fringe, where rivers deliver terrestrial sediments to the coast, and the orange shading indicates the arid coasts and interior, which are largely devoid of rivers and dominated by carbonate sediments.*

whether they are terrestrial quartz and/or marine-derived carbonate detritus. The eastern side of the continent is dominated by quartz sand, up to 98 per cent in southeast Queensland. There are two reasons for this. First, this is the humid section of the continent, where numerous rivers deliver usually quartz-rich sand to the coast. This is complemented by relatively few local sources of marine organisms. Even along the east Queensland coast, in lee of the Great Barrier Reef, the proportion of carbonate remains at 15 per cent, because the reef lies 10 to

100 kilometres offshore, too distant for the reef-derived carbonate material to reach the shore (as discussed in Chapter 7).

Carbonate-rich sediments dominate across the southern and up the western Australian coast, and contribute 50–60 per cent of beach sediments in the Kimberley and Northern Territory. A number of factors contribute to this dominance. First, in the south and west aridity results in few, if any, rivers and consequently little terrigenous sediment. Second, in the northwest, while there are several large rivers that deliver huge volumes of sediment to the coast, the low waves are insufficient to transport the sediment back to shore. This sediment remains in the subtidal and on the inner continental shelf. Third, these same regions have rich marine ecosystems to supply carbonate sand, including the northern fringing coral reefs and organisms living in the tidal sand and mud flats, which are located adjacent to the backing beaches. These regions also have tropical and particularly temperate seagrass meadows, which lie in shallow water adjacent to the shoreline, and the southern temperate carbonate shelf sediments, which although located in deep water up to kilometres offshore, are transported to the shore by the high waves of the Southern Ocean. As a consequence, approximately half of Australia's beach and dune sands are composed of calcium carbonate derived from marine organisms.

Sediment transport

Sediment transport is a function of the nature, level and direction of wave, tide, river or wind energy, and the size of the material (sediment) being transported. The role of sediment size is critical (see box 'Sediment size and transport', below).

Sediment size is highly relevant to sediment transportation by waves, currents and wind, and as a consequence most sediments on the open coast are graded by their transportability. Boulders, cobbles and coarse sand (> 1 millimetres) are so heavy they can only be moved as bedload along the bed of a river, inlet or sea floor, and subsequently they move very slowly and usually not very far. Medium to fine sand can be moved as bedload, and also in suspension for a few seconds to a minute or so, by strong currents, particularly in the presence of waves. Sand can therefore be transported slowly as bedload and more rapidly in interrupted suspension. Over time it can be transported long distances by river, waves, tides and strong winds. Sand is eventually deposited in river channels and estuaries and at the coast at river mouths and in deltas. The latter may be reworked along the coast by waves and tidal currents. Very fine sand, silt and clay are so fine and light that they can be lifted and transported in suspension at the speed of the river or currents. Consequently, they can be transported rapidly over long distances to settle out in quiet locations such as swamps, tidal flats, estuaries and the deep ocean.

Along Australia's wave-dominated shorelines there is sufficient wave energy to move cobbles and boulders very slowly across a short distance, resulting in only a few hundred cobble and boulder beaches. Large quantities of sand have been moved more rapidly to build more than 10 000 beaches and backing coastal dunes. Silt and clay are transported to quiescent locations and deposited on tidal flats at the shore. Around the Australian coast the vast majority of the beach systems are composed of sand

SEDIMENT SIZE AND TRANSPORT

Sediment is a material that has been eroded and transported by gravity, wind, water or ice; it includes mud, sand, gravel[1], boulders and organic debris. It can be described in qualitative and quantitative terms, as indicated below.

	diameter (mm)	settling velocity m/sec	transport mechanism water	air
Boulder	256	>1	bedload	stationary
Cobble[1]	64–256	>1	bedload	stationary
Pebble	4–64	>1	bedload	stationary
Granule	2–4	>1	bedload	stationary
Sand				
coarse	0.5–2	0.2	suspension	stationary
medium	0.25–0.5	0.04	suspension	traction/saltation
fine	0.0625–0.25	0.003	suspension	traction/saltation
Mud				
silt	0.0078–0.0625	0.00004	suspension	suspension
clay	0.00195–0.0078	0.000003	suspension	suspension

[1] The common term gravel includes the granule–pebble–cobble range of sizes.

Sediments range in size from clay to boulders, each within a standard range of diameters. Based on their size they have a settling velocity in water (the time it takes that size of sediment to settle in a column of water 1 metre deep) that exponentially decreases with decreasing size. The result is that in flowing water, granules and coarser materials move slowly along the bottom as bedload, while sand can go into limited suspension and mud into long-term suspension travelling at the speed of the flow. In air, coarse sand and coarser sediments are too heavy to be blown and thus remain stationary, while fine to medium sand can be rolled along the ground (traction), or go into limited suspension (saltation), while clays and silts in suspension form dust storms.

ranging from 0.1 to 0.6 millimetres – that is, fine to medium sand. This is because there is an abundant supply of such sand, transported to the coast by rivers and waves, and once there it remains within the turbulent surf zone. Finer silts and clays are trapped in sheltered estuaries or are flushed seaward to settle as mud on the continental shelf and deep ocean. Boulders and cobbles are difficult to move – they tend to stay close to their source, usually an eroding headland or river supplying gravel sediment.

Onshore, offshore and longshore transportation

Once sediment is delivered to the coast by rivers, waves or shoreline erosion, it may be transported in one of three directions: onshore, offshore or longshore. The sediment is also part of a littoral, or coastal, sediment cell, to which sediment may be added or lost. This affects the sediment budget and, as a consequence, the stability of the shoreline.

Onshore sediment transport is primarily driven by waves, with the sediment deposited as a beach, then possibly blown further onshore by wind to build coastal dunes. It is also moved by waves and tidal currents to be deposited in tidal inlets and tidal deltas. Sand deposited in dunes and tidal deltas may be permanently lost from the system, resulting in a net loss of sediment.

Offshore transport is driven by high-wave conditions, which can generate strong seaward currents and which may deposit the sand so far offshore it is lost from the system. Longshore transport relates to the movement of sediment along the coast, particularly in the surf zone and along beaches. Sediment moves along the coast until it reaches a sink: that is, a place of permanent removal

from the transportation system into a beach, dune, tidal delta or offshore. Headlands partition the coast into separate compartments, and there is usually little movement of sand around prominent headlands or across major inlets.

Regional sediment transport

At a regional scale there are several large areas of sediment accumulation, each located at the end of a major onshore or longshore sediment transportation system. Australia's coastal sediment systems are driven by the predominantly southerly Southern Ocean swell around the southern half of the continent, and by southeast trade-wind waves, assisted in parts of northern Australia by northward-flowing tidal currents. Figure 1.10 illustrates the major systems. The main conduit for longshore sand transportation takes place in the most energetic part of the beach: the surf zone, and particularly in the inner-surf zone, where longshore currents are strongest. Lesser amounts are moved seaward of the surf zone, and very little is moved in the swash zone (where waves wash up and down the beach). Seaward of the surf zone, north-trending tidal currents assisted by sediment entrained by wave action also contribute to the transport of sediment, while in northern Australia the higher tide ranges and strong tidal currents move large quantities of mud in suspension.

Cape York Peninsula

Cape York and particularly Torres Strait are the focus for two converging sediment transportation systems, both driven by southeast trade-wind waves and assisted by north-trending flood tidal currents,

Figure 1.10 *Major Quaternary longshore sand transportation and depositional systems. The orange arrows indicate major longshore sand transportation pathways; the blue arrows show dominant southerly swell direction; the green arrows dominant trade-wind wave direction; and the shaded areas indicate major regional sediment accumulation in Torres Strait, the southeast Queensland sand islands, the Murravian Gulf and Zuytdorp Cliffs, with appropriate volumes indicated in cubic metres.*

Figure 1.11 *Sand moving northward along the eastern Cape York coast has built the shoreline at Claremont Point 1.5 km seaward as a series of up to 20 beach-ridge spits and swales. Sand in this picture continues to move north around the point as recurved sand spits (foreground).* Photo: A.D. Short

particularly on the east coast. The sand is supplied by the numerous rivers on both sides of the cape. On the east coast, the northward transport is evident all the way up the coast (see Figure 1.11) from as far as Princess Charlotte Bay, 500 kilometres south of the cape. It is manifest in the north-trending spits and creek-river mouths, in northward migrating tidal sand waves and in the northwest orientation of the major coastal dune systems. The series of long, north-trending spits and recurved spits are best developed at the mouths of the Burdekin, Don and Elliott rivers (see Figure 1.12). The transportation

of sediment is interrupted by some of the major bays, such as Lloyd and Temple bays, but the sand gradually makes its way north, finally to accumulate in a massive tidal delta system, particularly on the western side of Torres Strait.

On the west coast of Cape York Peninsula, the movement of sediment is also evident in the northward displacement of creek and river mouths. A smaller, secondary system operates on the western shores of the Gulf of Carpentaria, as the trade-wind waves move sediment delivered by the Limmen Bight, Roper and other rivers north along the east

Figure 1.12 This 4-km-long, north-trending spit, with older spits at its base, and another spit to the north, is located at the Burdekin River delta. To the far right is the base of Cape Bowling Green, a major spit that extends 22 km to the north, while the low, mangrove-fringed delta occupies the top half of the image. Photo: A.D. Short

Arnhem Land coast, again in a system interrupted by several large bays.

Queensland sand islands

The three largest areas of sediment accumulation are all located around the high-energy southern half of the continent. The largest is that which extends along the northern New South Wales–southeastern Queensland coast, terminating in the world's largest accumulation of sand in six sand islands: Fraser, Cooloola, Bribie, Moreton, North and South Stradbroke. The almost-pure quartz sand that forms the islands was originally eroded from extensive granite rocks in the New England Highlands of New

South Wales and delivered to the northern New South Wales coast via the Hastings, Macleay, Richmond, Clarence, Brunswick and Tweed rivers, then transported by waves north along the coast from at least as far south as Port Macquarie, a total distance of 800 kilometres. Much of the sand transportation probably occurred at lower sea levels, when there were fewer of the headlands that presently compartmentalise the coast and interrupt the movement. It is estimated today that up to 500 000 cubic metres of sand is moving across the New South Wales border at the Tweed River onto Queensland's Gold Coast each year. As the six sand islands total about 250 cubic kilometres in volume, it would take – at the

Figure 1.13
Transgressive sand dunes moving towards Cape Howe (centre foreground), which marks the Victoria–New South Wales border and the northern terminus for a littoral sediment cell that commences at Wilsons Promontory and ultimately delivers sand to New South Wales (right foreground).
Photo: A.D. Short

present rate of supply – about 500 000 years to deposit. Given that the islands date back 2 million years, with sand moving northward throughout this period, this is more than an adequate rate of supply to build the islands. Today, the final terminus for this sand is the 30-kilometre-long Breaksea Spit, at the northern tip of Fraser Island. The island and spit have extended so far across the continental shelf that today the sand moves along the spit to the edge of the continental shelf, then cascades down the continental slope and is lost to the deep ocean.

South of Port Macquarie the northerly transportation is interrupted and essentially stopped by the numerous headlands that compartmentalise

the beaches, particularly south of Newcastle. Few rivers presently supply sand to the coastal systems along the south coast, but instead this sand is deposited in the estuaries and coastal lakes.

Eastern Victoria

Eastern Victoria has a separate longshore transport system that extends from Corner Inlet on the eastern side of Wilsons Promontory. It travels the length of the 220-kilometre-long Ninety Mile Beach, where it is interrupted by rocky sections and headlands of the Croajingalong coast and finally terminates at Cape Howe, a distance of 380 kilometres (see Figure 1.13). The predominantly quartz sand is derived from the inner continental shelf, as no rivers are supplying sand at the present high sea level. This system not only supplies wave-driven sand along the shore, but also east-trending dunes that cut across headlands such as Rame Head and at Cape Howe, to deliver sand directly to New South Wales. This sand appears then to be transported offshore to a massive inner-shelf sand body, centred on Gabo Island. At lower sea levels, some may have moved north into the Disaster Bay region, where today the shoreline has built seaward 2 kilometres into the bay.

Southern Australia

Across southern Australia, the waves and winds arrive from the southwest and, in most cases, drive sand directly onshore to form high-energy beaches backed by northward trending dunes that move inland. The largest accumulation of sand is located in the now largely infilled Murravian Gulf (see Figure 1.8). Here, onshore sediment transportation began building beaches and barriers over 2.5 million years ago and as much as 500 kilometres inland. The terrigenous quartz sand is sourced from the Murray–Darling river system and carbonate sand is derived from the inner-continental shelf, making a mixed quartz–carbonate province. Continued onshore sediment transportation, assisted by slight uplift of the shore, results in a 150 000-square-kilometre area of sand accumulation, containing approximately 200 cubic kilometres of sand.

Smaller-scale systems operate in both St Vincent and Spencer Gulfs, where transport into the gulfs results in several north-trending sand spits and barriers. These include Adelaide's Lefevre Peninsula, Webling Point at Port Broughton, Point Davenport and Germein Point at Cowell.

Western Australia

Once around Cape Leeuwin, the north-trending Western Australian shore is exposed to the prevailing southerly waves and, at times strong, southerly winds. Both these drive a northward sediment transportation system that operates north from Geographe Bay, with minor interruptions, all the way to Shark Bay. This sand is sourced primarily from carbonate production on the inner-continental shelf and in nearshore seagrass meadows. It averages 70 per cent carbonate, with some sections containing 99 per cent carbonate material. The areas of greater quartz sands occur close to river mouths such as those of the Swan, Gregory and Murchison, where the sand was delivered at lower stands of the sea. The ultimate sink for these sands are the northern shores, as manifest in the 300-kilometre-long, up to 200-metre-high, Zuytdorp Cliffs, which are similar in size to their counterparts on the southeast Queensland coast, and which hold approximately 200 cubic kilometres of largely carbonate sand.

Secondary and smaller, north-trending systems also operate along the carbonate-rich Ningaloo coast including north-trending dunes; along parts of the Pilbara coast in association with the numerous sandy river mouths; along Eighty Mile Beach and into Roebuck Bay (see Figure 1.14); and the eastern Dampier Peninsula north from Broome. In all these situations the transportation is evident in north-trending spits, often north-trending dunes, and occasional large sinks of sand in coastal bays.

Estimating the amount of sand being moved alongshore in any of the above systems is extremely difficult, owing to the large number of factors that contribute to the transportation of sediment: the fact that it can reverse periodically; the numerous sinks, some not always apparent, that can siphon off sand; and the attrition of carbonate sands through abrasion. The only credible estimate is that of 500 000 cubic metres per year for the northern New South Wales–southeastern Queensland system. There have been few estimates for any of the other systems. The total amounts of sand accumulated in the shaded areas on Figure 1.10 have been estimated at 250 cubic kilometres for southeast Queensland and islands, and 200 cubic kilometres for each of the Murravian Gulf and Zuytdorp Cliffs, all huge volumes of sand and of similar order of magnitude.

Conclusion

The Australian coast is the result of a series of geological events dating back 3500 million years. Around 1900–1700 million years ago, ancient cratons merged to form the Australian Craton, covering all of western and much of central Australia. Subsequently, and particularly between 500–50 million years, a series of mountain-building events successively built the eastern third of the continent. By 100 million years ago, the continent was separated from Gondwanaland and had begun its slow northward drift, defining the west and south coasts, as well as opening the Indian–Southern oceans. Subsequent rifting 100–50 million years ago opened up the Tasman–Coral seas and formed the east coast. Buckling around the edges resulted in a series of sedimentary basins, while continuing erosion of the continent continued to supply sediments to fill the basins and form sedimentary deposits around the coast. Finally, climate-controlled fluctuations in sea levels have alternately flooded the coast and

Figure 1.14 *A series of multiple recurved spits at Sandy Point enclose a mangrove-filled inlet and illustrate the northward (right to left) movement of sand into the southern section of Roebuck Bay, Western Australia.* Photo: A.D. Short

exposed the continental shelf, and the geologically most recent suite of landforms formed in the past few millennia as the coast was inundated by the most recent rise in sea level.

The present Holocene coastline comprises coastal sedimentary deposits, in places reflecting the legacy of landforms formed during previous Quaternary highstands. It consists of beaches, dunes, estuaries and deltas, usually separated by rocky sections. The sediments are derived from land-based terrigenous material and marine-derived carbonates. The predominantly sandy sediments have in some places been transported both onshore and longshore to form some of the world's major coastal beach and dune deposits.

COASTAL PROCESSES

THE GEOLOGICAL EVOLUTION of the Australian coast, as described in Chapter 1, is driven by the Earth's internal forces, which are manifested at the surface through plate tectonics as earthquakes, volcanoes and tsunami. The terrain produced by the Earth's movement is then shaped by the atmospheric forces of weather and climate, and at the coast by marine processes. The Australian coast is exposed to a range of tropical to temperate – and humid to arid – climates, with a similarly wide range of atmospheric and marine processes. This chapter examines the climate of Australia, along with coastal processes and how they both contribute to shaping Australian coastal landforms.

The climate of Australia

Climate is the long-term average of daily, monthly and seasonal weather. Climate includes the range of temperature, the type and amount of precipitation (rain, hail, snow), winds and extreme or unusual events, such as tropical cyclones and hailstorms. Climate varies from place to place across the Earth's surface because of variation in the amount of the sun's energy, or solar radiation. More solar radiation is received at the equator than at the Poles, producing a latitudinal temperature gradient from the hot tropics to the cold Poles. In addition, because the earth is tilted, more solar energy is received in summer – December in the southern hemisphere – than in winter, and this produces the seasons. Finally, because the Earth rotates every 24 hours, only the side of the Earth facing the sun receives solar radiation – hence we have bright, warm days followed by dark, cooler nights.

The sun produces both short wavelength radiation, which travels as light, and long-wave radiation as heat. The long waves

directly heat the Earth's surface and are reflected back into the atmosphere, where some are trapped in the clouds and thus warm the lower atmosphere. Once the surface is heated, the warmest equatorial locations produce hot air that rises and forms a low-pressure system, termed the 'equatorial low'. The rising equatorial air cools at elevation and descends to the Earth's surface at around 30° north and south of the equator, forming areas of cool, dry and clear air and these are called 'sub-tropical high-pressure systems'. Northern Australia is dominated in summer by the equatorial low, which draws in the hot, humid northwest monsoons, whereas during winter the subtropical high brings clear, dry conditions. At the Poles, the cold air is dense and sinks, producing a polar high. The out-flowing air from the polar highs and subtropical highs meet at an intense convergence zone around 60 degrees north and south of the equator where, warmed by the Earth's surface and oceans, they produce westerly-moving, upward-spiralling air known as the sub-polar low-pressure systems, also called 'mid-latitude cyclones'. The centre of Australia is under a subtropical high throughout the year, while in the south the cool, sub-polar low-pressure systems usually track south of the continent and are accompanied by strong westerly winds, which bring rain to the coastal fringe during winter.

Rainfall

Australia is a large, low continent with few mountains. It is entirely surrounded by seas and oceans, which provide a range of potential sources of moisture and precipitation. However, the dry subtropical high dominates much of the continent and, in order to induce rainfall, the moist maritime air masses must be driven onshore and preferably over high mountains such as the Great Dividing Range. The lack of mountains provides few opportunities for such orographic-induced rainfall and, as a result, potential rain-bearing air masses pass over the low, uninterrupted terrain. Only during summer do the northwest monsoons and southeast trade winds move moisture inland in the north and bring the Wet, while during winter the westerlies bring cold fronts and associated frontal rain to the south of the continent. Much of the interior remains arid throughout the year, because it is low and flat, dominated by the high and far from the coast and sources of moisture.

Australia's climate is therefore a product of its latitudinal position (9–43 degrees South) and associated pressure systems, its continental size, low elevation and surrounding oceans. It can be classified into six simplified climate zones (see Figure 2.1). They range from the hot, humid equatorial-tropical systems of the north, and the arid-to-semi-arid centre, to the more humid, subtropical and temperate systems in the south, with all six zones impinging on parts of the coast.

Temperature

The distribution of temperature in Australia reflects latitude, proximity to the coast and elevation (see Figure 2.1). Latitude controls the amount of solar radiation and results in the highest mean maximum temperatures occurring in the tropical north, where they average more than 33 degrees Celsius in summer, decreasing south with latitude to as low as 12 degrees Celsius in Tasmania, with cooler

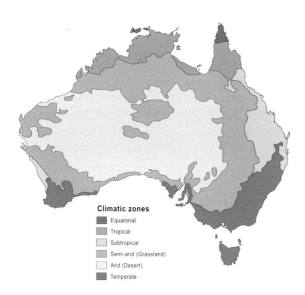

Climatic zones
- Equatorial
- Tropical
- Subtropical
- Semi-arid (Grassland)
- Arid (Desert)
- Temperate

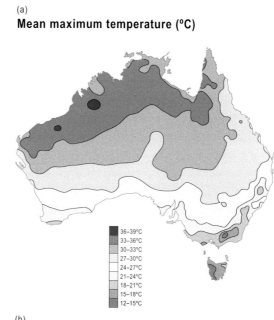

(a)
Mean maximum temperature (°C)

36–39°C
33–36°C
30–33°C
27–30°C
24–27°C
21–24°C
18–21°C
15–18°C
12–15°C

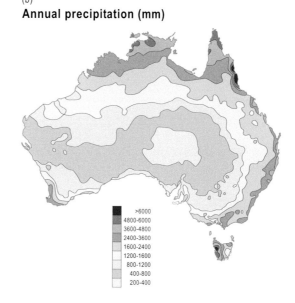

(b)
Annual precipitation (mm)

>6000
4800-6000
3600-4800
2400-3600
1600-2400
1200-1600
800-1200
400-800
200-400

Figure 2.1 Australia's climatic zones (left); *a)* mean maximum temperatures (right), which are predominantly latitudinally controlled; *b)* annual precipitation (rainfall, snow etc.), which has a summer maximum in the north, a winter maximum in the south and an arid interior and west coast.

temperatures also in the elevated Australian Alps. The temperatures are moderated along the coast, particularly along the east coast, where cooler, humid easterlies flow in from the sea to reduce temperature by a few degrees.

Precipitation

Precipitation is derived from moisture taken up by evaporation over the oceans and some tropical lowlands, and transferred to the continents by onshore winds. Once over land it may fall as convectional rain; that is, hot, moist rising air, especially in the tropics; as orographic rain where the moist air is made to rise over a mountain range; or as frontal rain, where moist, cold fronts produced by sub-polar lows rise up over denser air. In all cases, the rising air causes water vapour to condense, forming clouds and eventually falling as rain, snow, sleet or even hail. Most precipitation around Australia falls as rain, apart from sleet and snow in the Australian Alps, and occasional hailstorms. Figure 2.1b indicates the humid coastal fringe and arid interior. The northern humid fringe receives its rain from the northwest monsoon between November and April, together with rain associated with the southeast trade winds along the eastern Queensland coast. This rain is periodically supplemented by torrential downpours during tropical cyclones. It reaches a maximum up to 9600 millimetres in the Cairns region and over 4800 millimetres along the coast of the Northern Territory and Cape York Peninsula. In the north it decreases inland, while along the east coast much of the rain falls along the eastern side of the Great Dividing Range, with a drier rain shadow zone extending west of the ranges (see Figure 2.1).

The pressure and wind systems that produce this climate are discussed below, beginning in the north and working south. This is followed by a discussion of Australia's coastal climate.

Pressure systems and winds

Equatorial lows

The distribution of pressure systems and the Coriolis effect (see box 'The Coriolis effect', below) give rise to the major wind patterns. These systems include:

- polar easterlies that spiral outwards as a low-velocity, easterly flow around the polar highs
- a zone of strong westerly winds that rings both hemispheres centred on 60 degrees latitude associated with the sub-polar lows
- the trade winds that blow between the subtropical high and the equatorial low and produce a wide zone of moderate-velocity winds blowing from the northeast in the northern hemisphere and southeast in the southern hemisphere
- at the equator, the rising air of the equatorial low pressure produces the calm doldrums.

When the trade winds cross the equator in the summer, they are deflected to become the southwest monsoon in Asia and the summer northwest monsoon (the 'Wet') across northern Australia.

The equatorial low shifts to the southern hemisphere from November to April and sits over northern Australia, centred on the Pilbara and Cloncurry, areas of low pressure generated by the intense summer heat. The hot, rising air is replaced by a northward flow of warm, humid air arriving from the tropics. This air is known as the northwest monsoon, which prevails from November to April and brings warm, wet conditions across the north between Broome and Cape York Peninsula. The associated low-velocity winds also generate low, short waves on west and north-facing shores across the north.

The Intertropical Convergence Zone

The northwest monsoon winds from the north, and the southeast trade winds from the south, converge with the air from both systems rising under the equatorial low. This zone is known as the Intertropical Convergence Zone (see Figure 2.2), which is a highly dynamic zone with masses of warm, humid air rising to great heights. These tropical areas are characterised by intense monsoonal precipitation during the wet season, often as afternoon thunderstorms. It is also a zone where the colliding northwest and southeast winds can cause disturbances, which spiral around each other to form a tropical depression that may deepen to become a tropical cyclone. In the Australian region, tropical cyclones originate under the Intertropical Convergence Zone over the warm, humid ocean regions, in the eastern Indian Ocean off northwest Australia, in the Gulf of Carpentaria and the Coral Sea off the northeast coast.

Cyclones

Cyclones are low-pressure systems, also called 'lows', which form over oceans and have inward-spiralling, moist rising air. They rotate clockwise in the southern hemisphere and comprise:

- *tropical cyclones* that form in the tropics during summer, usually between 10–20 degrees latitude and are the most intense form of cyclone, known as 'hurricanes' in the Americas and 'typhoons' in Asia
- *east-coast cyclones* that form occasionally off the east coast of Australia, centred on 30–35 degrees latitude
- *mid-latitude cyclones*, also know as sub-polar lows, that form throughout the year over the Southern Ocean, centred on 60 degrees South.

Tropical cyclones form over the oceans between 10–20 degrees North or South (see Figure 2.3a). They cannot form within 5 degrees North or South of the equator because the Coriolis effect reduces to zero at the equator, meaning that there is no force to cause the winds to be deflected and spiral to form a cyclone. They form over the warm tropical oceans because they require the rising heat and, in

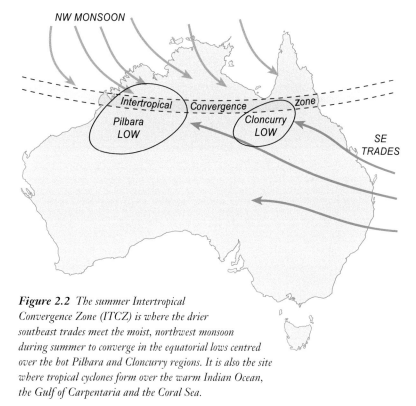

Figure 2.2 *The summer Intertropical Convergence Zone (ITCZ) is where the drier southeast trades meet the moist, northwest monsoon during summer to converge in the equatorial lows centred over the hot Pilbara and Cloncurry regions. It is also the site where tropical cyclones form over the warm Indian Ocean, the Gulf of Carpentaria and the Coral Sea.*

THE CORIOLIS EFFECT

Air flows from areas of high pressure to areas of low pressure, and this produces the global wind systems; the greater the pressure difference between the high and low, the stronger the wind. The wind is also diverted by a force called the Coriolis effect, exerted on any object (gas or water) moving over the Earth's surface owing to the rotation of the Earth. The Coriolis effect means that an object will be deflected to the right in the northern hemisphere and to the left in the southern hemisphere. It is the Coriolis effect that causes cyclones to rotate counter-clockwise in the northern hemisphere and clockwise in the southern hemisphere. The force is greatest at the Poles, decreasing to zero at the equator. The Coriolis effect influences the paths of tropical cyclones, causing them to veer south (that is, to the left) once they have formed to the north of Australia.

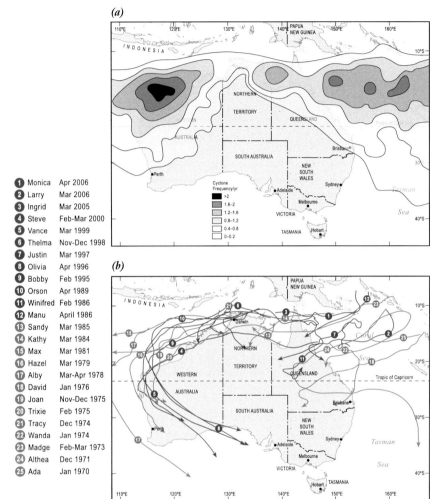

1 Monica   Apr 2006
2 Larry   Mar 2006
3 Ingrid   Mar 2005
4 Steve   Feb-Mar 2000
5 Vance   Mar 1999
6 Thelma   Nov-Dec 1998
7 Justin   Mar 1997
8 Olivia   Apr 1996
9 Bobby   Feb 1995
10 Orson   Apr 1989
11 Winifred   Feb 1986
12 Manu   April 1986
13 Sandy   Mar 1985
14 Kathy   Mar 1984
15 Max   Mar 1981
16 Hazel   Mar 1979
17 Alby   Mar-Apr 1978
18 David   Jan 1976
19 Joan   Nov-Dec 1975
20 Trixie   Feb 1975
21 Tracy   Dec 1974
22 Wanda   Jan 1974
23 Madge   Feb-Mar 1973
24 Althea   Dec 1971
25 Ada   Jan 1970

Figure 2.3 a) *The location and frequency of summer tropical cyclone formation across northern Australia; b) the trajectory of some of the more damaging tropical cyclones since 1970. Notice how most trend west and south in a counter-clockwise trajectory, and usually dissipate when they move over land.*

particular, water vapour to draw their energy. As the water vapour rises it condenses and in doing so releases latent heat of condensation, which provides both the energy to drive the cyclone and the intense precipitation associated with the cyclone. Tropical cyclones therefore form during summer over tropical seas where the sea surface temperature is greater than 26 degrees Celsius, resulting in strong evaporation rates and associated high humidity. Their formation is also aided by calm wind conditions, with the initial spiralling winds deflected to the left (in the southern hemisphere) by the Coriolis effect (see box, above).

Categories of tropical cyclones

Once formed, tropical cyclones tend to move west, under the influence of the southeast trade winds, and veer south because of the Coriolis effect (see Figure 2.3b), although their behaviour in the Australian region is more erratic than in other parts of the world where they are experienced. This southwesterly trajectory means that many move down the northwest coast of Western Australia, where increasing deflection can cause then to make landfall, particularly in the Pilbara region. In the Gulf of Carpentaria they tend to make landfall in the southern gulf, while in the Coral Sea they tend to head southwest towards the Queensland coast. Tropical cyclones in which the maximum wind gust is up to 125 kilometres per hour are called 'category 1'; those with gusts up to 165 kilometres per hour are 'category 2', which can cause damage to buildings. There are three categories of severe tropical cyclone: category 3, with gusts of 165–225 kilometres per hour, category 4 with gusts of 225–280 kilometres per hour, and the most destructive, category 5, with gusts in excess of 280 kilometres per hour. Once they move over land, tropical cyclones are cut off from their moist energy source and usually degrade rapidly into a tropical depression.

Cyclone-prone locations

Tropical cyclones form between November and April, with a pronounced late-summer peak when sea conditions are hottest (see Figure 2.4b). They have been experienced along the entire northern Australian coast (see Figure 2.3), but are most prevalent off northwest Western Australia, where they tend to make landfall between Port Hedland and Onslow, centred on 20 degrees South (see Figure 2.4a), in the Gulf of Carpentaria, where they tend to cross the southern Gulf coast, and in the Coral Sea, where they can land between Cape York and Brisbane. Of these, 40 per cent cross north of Cooktown and 50 per cent land between Cooktown and Mackay. Most of the Kimberley, Northern Territory and northern Cape York Peninsula experience fewer cyclones because all of the northern coast lies between 9 and 18 degrees South, while the cyclones tend to be generated between 15 and 20 degrees South, making landfall further south and thereby usually missing much of the northern coast.

Super cyclones

When tropical cyclones do make landfall, they are accompanied by heavy rainfall and very strong winds that generate big seas and storm surges, all of which can have a devastating impact on the coast through wind damage (see box 'Cyclone Tracy, right'), river and storm-surge flooding, and wave erosion and run-up. Although tropical cyclones have a low frequency of occurrence and impact at any particular location, the imprint can persist for millennia as stranded,

CYCLONE TRACY

Cyclone Tracy devastated Darwin on Christmas Day in 1974. The cyclone originated over the Arafura Sea on 20 December, hit Darwin early on the morning of 25 December and caused widespread damage. Wind speeds of 110 kilometres per hour, gusting to 195 kilometres per hour, were recorded before the anemometer was wrecked by a gust of 217 kilometres per hour. The eye of the storm was 12 kilometres across and it passed within 8 kilometres of Darwin airport. Buildings were flattened and 49 people lost their lives, with a further 16 reported lost at sea. Storms of this magnitude have a probable annual return period of around 100 years; powerful storms in that part of the Top End occurred also in 1827, 1839, 1897, 1915, 1919 and 1937.

storm-generated coastal features including cheniers, shingle and boulder ridges, and other elevated storm deposits. The size and elevation of some boulder ridges across northern Australia suggest that even bigger tropical cyclones have occurred in the past than those observed since European settlement. These 'super-cyclones' and the storm surges they have generated are likely to pose a major threat to coastal settlements if they recur in the future.

Subtropical high-pressure system

The subtropical high-pressure system influences Australia's climate throughout the year. During summer it is centred on 32 degrees South, covering the central–southern part of the continent, where it brings clear skies and hot, dry summer conditions (see Figure 2.5). On the eastern Queensland coast, the high draws in the southeast trade winds to deliver a broad band of summer rain to coastal Queensland and northern New South Wales. On the south side of the high, the humid, sub-polar low-pressure systems are kept well south of the continent, rarely reaching the coast, except in Tasmania.

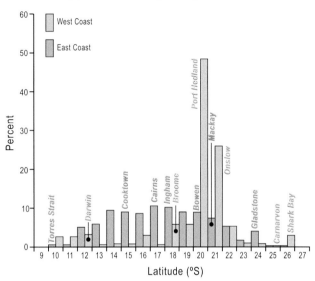

(a) Australian tropical cyclone landfalls

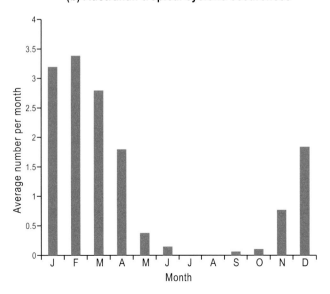

(b) Australian tropical cyclone occurences

During winter, the equatorial low moves northwards over south Asia and the high pressure shifts north to be centred at 30 degrees South and, in this position, dominates the entire Australian continent. Winds blow anticlockwise around the high, blowing as moderate-velocity, southeast trade winds across the north, where they bring limited winter precipitation to eastern Queensland. They also blow as a cool, dry wind across the remainder of the Top End, bringing dry winter conditions across the north (see Figure 2.5). The trade winds also generate moderate seas on all east and south-facing coasts.

Sub-polar lows

In the south, the mid-latitude cyclones move closer to the continent in winter, with occasional lows and associated southerly winds and cold fronts reaching the coast and bringing associated winter rains. They also generate a strong westerly flow known as the 'roaring forties', 'raging fifties' and 'screaming sixties', depending on the latitude.

In southeast Australia, a fourth regional pressure system, east coast cyclones, has a major impact on the southeast Australian coast between Fraser Island and Tasmania, particularly on the

Figure 2.4 a) *Latitude of tropical cyclone landfalls in Western Australia (green) and Queensland (yellow). Note the peak in the west in the Pilbara region (20 degrees South) of Western Australia, the greater spread down the Queensland coast, and the fact that none land north of 10 degrees South; **b)** Average number of tropical cyclones per month forming off northern Australia. Note the late summer January–March maximum, when sea temperatures are highest, and none in the cooler mid-winter months.*

(a)

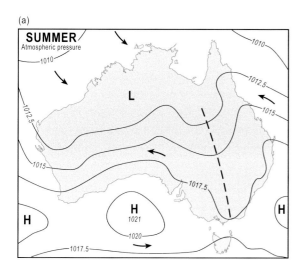

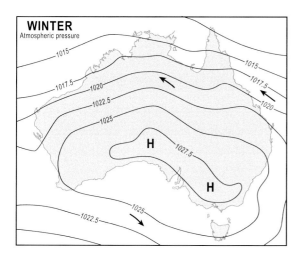

(b)

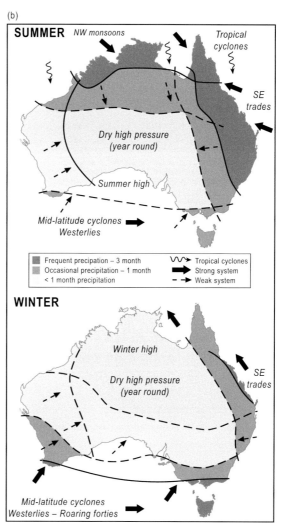

Figure 2.5 a) *Australian mean atmospheric pressure in summer and winter. Note how the equatorial low dominates the north during summer, with the subtropical high located just south of the continent. During winter the low has moved to the northern hemisphere and the high dominates the entire continent. Arrows indicate generalised seasonal wind direction.*

b) *Generalised summer and winter pressure and wind systems impacting Australia, and their degree of penetration and delivery of potential rainfall. The north is dominated by summer tropical rains, while the south relies on winter rainfall (and snow) originating from the sub-polar lows.*

New South Wales coast. These intense lows can form throughout the year, although they are more prevalent in late summer–early winter, with 10 lows forming on average each year, although more than 20 form in some years. As they intensify over, and just off, the coast, they bring strong easterly winds, heavy rain and big seas to the southeast coast, potentially causing coastal flooding, wind damage and substantial, and occasionally severe, shoreline erosion. The strong winds and big seas that drove the ship *Pasha Bulka* onto a Newcastle beach in June 2007, in addition to causing widespread flooding in the Hunter Valley, were part of an intense east-coast cyclone.

Sea breezes

The final important coastal wind system is the local sea breeze, which occurs particularly during summer when the high-pressure system is centred over the continent. The high is usually accompanied by clear skies and light regional winds, which permit intense surface heating. Over the land, the hot air rises in the mid to late morning. At the coast, the rising air is replaced by cooler air moving in from the adjacent sea, resulting in a flow of cooler sea air by late morning – the sea breeze. Its direction is dependent on the orientation of the coast and the Coriolis effect, and ranges from northeast on the east coast to southwest on the west coast. Along the New South Wales coast, sea breezes occur on average 14 days a month during summer and 9 days a month during winter, when they also tend to be lighter. The strongest sea breezes (up to 45 kilometres per hour) occur along the southwest Western Australian coast, where in Perth their cooling effect is known as the 'Fremantle Doctor'.

Australia's coastal climate

In the coastal zone, the Australian climate directly controls rainfall, temperature and local winds. The climate of the ocean, known as 'physical oceanography', also has an effect on the coast. In this section we consider the impact of wind on the regional oceanography, as well as the role of ocean waves, tides and currents.

Coastal winds

Upwelling

Wind has three vital roles in the coastal zone: it drives ocean currents, generates ocean waves and blows beach sand inland to form sand dunes. Wind blowing over the ocean moves the surface, forming surface currents, which flow in the direction of wind. At the shore, offshore winds can cause the surface waters to move offshore, with the warmer surface water replaced by cooler water welling up from the seabed, and this can result in a sharp drop in water temperature.

Along the New South Wales coast, such upwelling is usually associated with moderate to strong, hot northerly winds. The northerly winds generate a south-flowing coastal current, which is deflected by the Coriolis effect to the left – that is, offshore. The cooler bottom water wells up close to shore, reaching the surface and sometimes causing a sudden 5–7° Celsius temperature decrease. Thus, paradoxically, the hottest days are often associated with some of the coldest ocean water.

Downwelling

Conversely, a strong southerly wind blowing along the New South Wales coast drives a northerly coastal current, which is also deflected towards the coast. Downwelling brings warm ocean water to the coast and raises water temperature by 1–2 degrees Celsius. So the cooler southerly is accompanied by above-average water temperature. These effects are illustrated in Figure 2.6, which shows the average sea temperature at Sydney, the seasonal range of temperature and the actual day-to-day temperature for one year.

Upwelling events are manifested by sudden drops in temperature, and downwelling by sudden spikes in temperature. The seasonal effect drives the gradual winter decline, reaching the lowest temperature in August, and the summer rise in temperature, which peaks in February. The lag in maximum and minimum temperatures is due to the fact that seawater warms and cools slower than the adjacent land surface.

Wind-blown sand transportation

The most enduring effect of wind on the coast is its ability to erode sand from the beach and transport it landward, to be deposited as coastal sand dunes. This wind-driven process is known as 'aeolian sand transport' (after Aeolus, the Greek god of wind). Because beaches are always bare of vegetation, the unconsolidated sand can be blown inland by strong winds to form sand dunes. Coastal dunes fringe the entire Australian coast, with the largest dunes all located in areas of high waves (that bring sand to the shore) and strong onshore winds that can blow large quantities of sand inland. These large dunes occur along the southern half of the west coast as a result of strong sea breezes, right across the south and parts of the southeast coast, where strong south to southwest winds prevail, and in the northeast, where seasonally strong southeast trade winds dominate. The role of wind in moving sand and forming Australia's very extensive – and in places massive – sand dunes is explored further in Chapter 6.

Over the sea, wind generates local wind waves, particularly those associated with the trade winds in the north and sea breezes right around the coast. These waves make a major contribution to the coastal wave climate, which is discussed in the following section.

Sydney ocean temperature

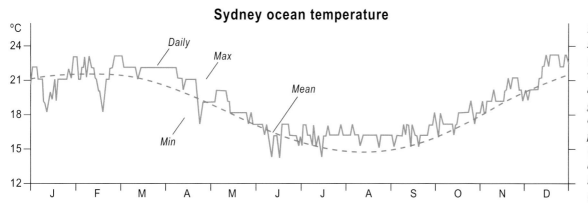

Figure 2.6 Mean sea surface temperature (dashed line) for Sydney, with a plot of daily changes in temperature generated by cooler periods of upwelling and warmer periods of downwelling (jagged line), all encapsulated by the range in mean temperatures (shaded area).

Wave climate

Ocean waves are the single most important process affecting the coast. They provide most of the energy to form and shape the shoreline. Ocean waves are a product of winds blowing over the surrounding oceans, and thus are related to atmospheric climate. Wave climates contribute the waves that break daily around the Australian coast, and provide most of the energy to drive coastal evolution. They are therefore highly relevant at a range of time scales.

Wave climate refers to the source of the waves as well as the seasonal variation in the height, period and direction of wave approach at a particular location. Waves are generated by wind blowing over the ocean surface and wave height and period (the time between successive waves) are directly related to four factors: wind velocity, wind duration, the fetch or length of ocean over which the wind blows and the depth of the ocean.

Wave energy

The highest waves and those with the longest periods are generated by strong winds that blow in a constant direction, for a long duration and over a long fetch of deep ocean. Wave energy is a function of the square of the wave height and wave period, meaning that a 10-metre-high wave has 100 times the energy of a 1-metre-high wave, and a 10-second wave has twice the energy of a 5-second wave. The height of a wave, and the amount of energy it contains, increases at the square of the wind velocity. Very strong winds are required to produce the highest waves (see Figure 2.7). Winds blowing long enough over an ocean will produce a fully arisen sea; that is, the highest and longest waves that can be produced by winds of that velocity.

Wave direction

Wave direction is a function of the wind direction, but is also diverted by the Coriolis effect. The world's biggest wave factories are therefore those where there are strong, persistent winds, blowing for long periods (days) over long stretches of the ocean (thousands of kilometres). While the most persistent winds and longest fetches are associated with the trade winds, they are of moderate velocity and rarely produce waves over 2 metres. The mid-latitude cyclones and their strong westerly winds blowing across the north Pacific and Atlantic oceans, and around the great Southern Ocean, produce the world's highest waves (see Figure 2.8). All these waves, because of their deflection by the Coriolis effect, head towards the equator, arriving at the coast as energetic southwest (southern hemisphere) and northwest (northern hemisphere) swell.

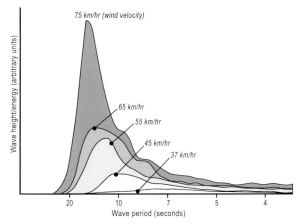

Figure 2.7 *As wind velocity increases, the period or time between waves (and their length) increases, and the amount of energy transferred to the waves increases exponentially. Note how, as wind velocity doubles from 37 to 75 kilometres per hour, the amount of energy increases exponentially. Very strong winds are required to produce large, high and long waves.*

Sea and swell

Ocean waves comprise sea and swell. Sea refers to waves that are still being generated by blowing wind; they tend to be shorter, higher, steeper and more irregular in height and direction. The crest of over-steepened seas can break in deep water. Swell refers to waves that have left the area of wave generation and travel under their own energy. Compared to sea, swell waves are lower, longer, less steep and more regular in height and direction.

Wave length, period and breaking

Wave length is the horizontal distance from crest to crest, whereas wave period is the time between two successive waves. These are related, in that length and speed are dependent upon wave period (speed increases proportional to period and length at the square of the period). In shallow water – that is, water whose depth is considerably less than the length of the wave – speed is controlled by water depth, slowing as shallower water is encountered.

As waves approach the shore they slow, shorten, steepen and tend to increase in height before ultimately breaking (see Figure 2.9). Breaking occurs when waves enter shallow water and steepen; the trough, beneath which water depth is the shallowest, slows down and is partially overtaken by the following crest, in the process of wave breaking. Waves break as spilling, plunging or surging breakers.

Local wave climates

All other winds also produce waves, some quite low and short, like those associated with the low-velocity, northwest monsoons, the polar easterlies and sea breezes, and some very high, like those associated

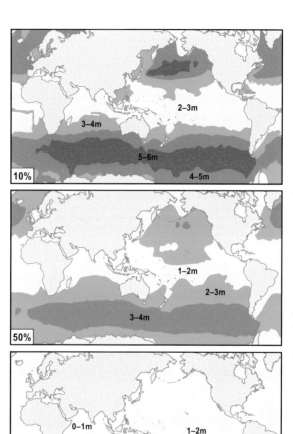

Figure 2.8 Average global deepwater wave height. The highest waves occur in the Southern Ocean, and north Pacific and the Atlantic, where they average 5–6 metres over 10 per cent of the year (top). The Southern Ocean has the most persistent higher waves, averaging 3–4 metres over 50 per cent of the time (centre), and are at least 2–3 metres high 90 per cent of the year (bottom). While the trade winds also blow year-round across huge sections of ocean, they are of only moderate velocity and hence generate only moderate waves (1–2 metres for 90 per cent of the year).

MEASURING WAVES

Wave height is the vertical distance from the wave's trough to its crest. Waves around the Australian coast are measured by a series of wave-rider buoys. Each buoy is anchored to the sea floor, usually in about 50–80-metre depth, 2 to 5 kilometres offshore, where they measure the deepwater wave height, period and direction. The buoys are fitted with an accelerometer, which senses the water movement and hence the wave height and period. This information is transmitted back to shore.

A series of buoys are located along most of the east coast, as well as off Tasmania's Cape Sorrell, at Cape Du Couedic on Kangaroo Island and off the Western Australian coast. Such wave data is essential for recording wave climate and the size of extreme waves, as well as for detecting changes in wave climate that may be induced by climate change.

Although surfers are particularly interested in wave conditions, they tend to grossly underestimate wave height, and use imperial measurements (that is, feet). The best way to visually estimate wave height while a surfer is on the wave is to assume the average, slightly crouched surfer is 1.5 metres high and to use that as a scale. Twice the height of a surfer equals 3 metres and so on.

Figure 2.9 *A long-period ocean swell shoaling, refracting and breaking around the entrance to Earlham Lagoon, Tasmania.*
Photo: A.D. Short

with intense tropical and east-coast cyclones. The combination of waves from all potential sources determines a location's wave climate. Some locations have simple wave climates that are dependent on only one source of waves from one direction, while others have more complex wave climates that have up to five wave sources from different directions. Australia's wave climate is related to the surrounding atmospheric pressure systems and associated wind regimes, and can be divided into a northern and southern wave climate, together with a southeast subsystem. These are briefly discussed in the following section.

Northern Australian wave climate

In northern Australia there are four potential sources of waves: the northwest monsoons, tropical cyclones, trade winds and sea breeze. The summer northwest monsoonal winds are low to moderate in velocity and generate low, short seas on west through to north-facing shores between November and April. In summer there are occasional tropical cyclones, which, while they can bring high and damaging seas, have a low frequency of occurrence and therefore only make a very minor contribution to the annual wave climate. During summer on the east coast, and during winter right across the Top End, the moderate to occasionally high-velocity southeast trade winds dominate and can blow for days at a time. They bring low to moderate seas to all east-facing shores, but blow offshore and bring calm conditions to west-facing shores.

In Queensland, the Great Barrier Reef intercepts the ocean seas, with only smaller trade *wind* waves generated within the Great Barrier Reef lagoon and reaching the mainland shore. Sea breezes can occur throughout the year, although they are more prevalent during summer. They tend to bring easterly winds and low seas on east-facing shores and westerly winds and low seas on west-facing shores, and are usually superimposed on one of the other wave systems described earlier.

The net result is a wave climate across the north that peaks at a moderately high, short sea (height 0.5–1.5 metres, wave period 3–6 seconds) in exposed locations, and in most locations less than this. Two additional factors decrease the already low seas at the coast. First is the low-gradient nearshore zone around the northern coast, which induces substantial wave shoaling and makes the already low waves even lower as they lose energy crossing these wide, shallow nearshore zones. Second is the orientation of the coast and numerous local bays, headlands, reefs and gulfs, which further limit wave generation and shelter many shores. The result along a heavily embayed coast, such as the Kimberley, is a breaker-wave height averaging only 0.1 metre, with the highest average waves only reaching 1 metre (see Figure 2.10). In the Northern Territory, waves are higher (average 0.3 metre) on the exposed east Arnhem Land coast, where they can reach a maximum of 1.5 metres. The highest average breaker-wave height is around Cape York Peninsula, particularly the east coast, where they average 0.4 metre and regularly reach 1.5 metres and higher during strong trade winds.

Southern Australian wave climate

The southern Australian coast receives persistent, moderate to high swell from the Southern Ocean.

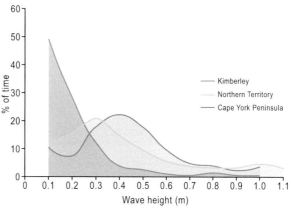

Figure 2.10 A plot showing regional breaker-wave heights across the northern Australian coast. Along the sheltered beaches of the Kimberley, most of the time waves are only 0.1 metre high, rarely reaching above 0.4 metre, while around the Territory coast and Cape York they are usually 0.3 metre and 0.4 metre high, respectively, and can reach 1 metre and higher.

The combination of persistent, strong westerly winds of relatively constant direction and velocity, with a long duration (days) and very long ocean fetch of great depth, enables waves to be generated that commonly reach several metres and average up to 4 metres (see Figure 2.8). At the coast their impact can be gauged from Figure 2.11, which presents the average monthly wave height and period for Cape Sorell (west coast of Tasmania), Cape Du Couedic (Kangaroo Island, South Australia) and off Perth.

At Cape Sorell, waves average 3 metres all year round, with a period of 12–13 seconds, an average daily maximum wave height of 5 metres and waves occasionally up to 18 metres. Low waves and calms are rare. These waves are generated by the continuous flow of mid-latitude cyclones centred on 60 degrees South. As the waves leave the generating area they become lower and longer, southwest swell bringing moderate to high wave conditions throughout the year across southern Australia, from North West Cape to Fraser Island.

Waves are biggest along the southern coast between the west coast of Tasmania and Cape

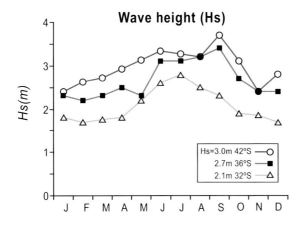

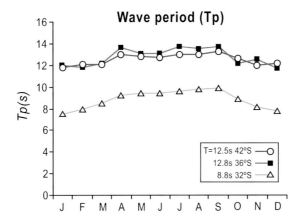

Figure 2.11 *Significant wave height (Hs) and peak period (Tp), as recorded by wave-rider buoys located at Cape Sorell, Cape Du Couedic and off Perth. Note the persistent high waves and long periods peaking during later winter, and gradual decrease in height to the north.*

Naturaliste, with wave height decreasing up the west coast, as indicated by the decreasing wave height between Cape Sorell at 42 degrees South, Kangaroo Island at 36 degrees South and Perth at 32 degrees South. Across the southern and southwest coast, these long swell conditions are augmented by summer sea breeze conditions, which generate short moderate seas, predominantly from the south.

The southeast coast between southeast Queensland and eastern Tasmania receives the Southern Ocean swell on average 200 days a year. As it moves up the coast, the swell has to bend and refract to reach the shore, as a consequence of which wave height is reduced to an average of 1.6 metres. Although this is the dominant wave, the coast also receives waves from four other sources: east-coast cyclones, tropical cyclones, high-pressure systems and sea breeze.

At Sydney, an average of 10 east-coast cyclones a year produce waves from the east on 55 days a year, with a mean height of 2.8 metres, the highest wave source on the east coast. Summer tropical cyclone waves arrive from the northeast on average 18 days a year. When highs are located off the coast they produce an easterly flow of air and low easterly waves, while northeast sea breeze waves arrive on average 13 days per month during summer. The result is a highly variable wave climate generated by three cyclonic sources (tropical, east coast and mid-latitude), the high-pressure system and local sea breezes, which vary in direction (northeast, east and south), seasonality, height and period. Figure 2.12 illustrates that as the sources of these waves vary, so too does their distribution along the coast, with southerly swell dominating to the south, whereas east-coast cyclones affect the entire coast more equally, and tropical cyclones have their greatest impact to the north. When all these sources are combined, deepwater waves average 1.6 metres at Sydney, with a period of 10 seconds, the dominant waves generated by the mid-latitude cyclones arriving from the southeast and secondary waves arriving from the east and northeast.

The west coast of Western Australia has two wave climates located either side of North West Cape. To the south, southwest swell generated by the mid-latitude cyclones dominates the coast from Cape Leeuwin up to North West Cape. This swell arrives year round, with a winter maximum, and averages 3 metres along the south coast, while decreasing to 2 metres up the west coast (see Figure 2.11). It reaches all exposed sections of shore, but is attenuated along the many sections of coast fronted by reefs and islands, and further north by coral reefs. The reefs substantially reduce wave height at the shore, as in the region between Cape Naturaliste and Kalbarri. Consequently, the breaker-wave height ranges from the full 2 to 3 metres on unprotected shores, to zero in sheltered locations. The swell is supplemented by local wind waves, which consist of strong southerly sea breeze waves in summer, and less persistent, though strong, northwest wind waves in winter. While these seas are superimposed on the swell on exposed sections, along many sections sheltered from the ocean swell, including the Perth coast, the wind waves can dominate the wave climate.

To the north of North West Cape–Exmouth Gulf, the coast trends northeast and does not receive the southerly swell. Further to the north, including the Pilbara and Kimberley coast, locally generated westerly wind waves and west to northwest summer

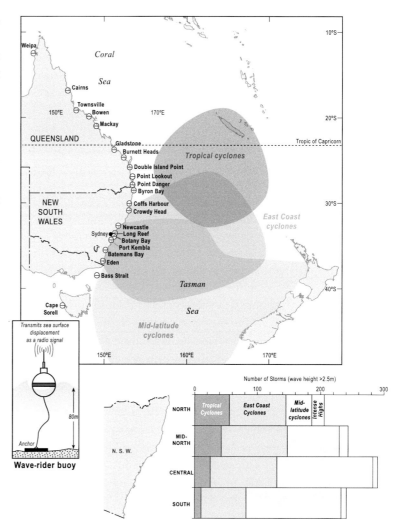

Figure 2.12 *Location of wave-rider buoys along the east Australian coast. The buoys are maintained by state government agencies in Queensland and New South Wales. This map also indicates the areas where tropical, east-coast and mid-latitude cyclones are located when their waves arrive at Sydney; and (lower) the occurrence of waves greater than 2.5 metres that are generated by these cyclones and intense highs along the New South Wales coast.*

sea breeze waves dominate the wave climate. These generate low, short seas and a lower-energy wave climate (see Figure 2.10), except during occasional tropical cyclones. During winter, lower-velocity westerly winds and offshore southeast trade winds result in low waves to calms at the shore.

Tides

Tides are the second major source of marine energy required to shape our coastal landforms. Unlike waves, which depend on winds for their generation, tides occur because the gravitational pull of the Moon and Sun, acting on a rotating Earth, causes a very slight bulge in the ocean surface. The Moon, being closer, contributes two-thirds of the tidal force and the more distant Sun the remaining one-third. The period, or time between the tides, is precise and can be predicted years in advance because the orbital characteristics of all three are quite precise. The Moon, or lunar, tide has a period of 12.42 hours, while the Sun tide occurs every 24 hours. This results in the two going in and out of phase on a 14 and 28-day cycle, the latter known as the 'lunar tidal cycle', or lunar month.

Tides are highest (spring tides) when the Moon and Sun are aligned and in phase, and when they combine their gravitational pull. Tides are lowest (neap tides) when they are at right angles and 90 degrees out of phase. On a perfect Earth covered in very deep water, the tide would rotate around the Earth, as a wave of low height (< 20 centimetres) but considerable length (hundreds of kilometres), every 12.42 hours, resulting in two tides a day, called 'semidiurnal tides' (that is, they occur twice a day). However, the Earth's surface has relatively shallow oceans that are divided into ocean basins by land, which occupies 25 per cent of the surface. In order to accommodate the tide in the shallow irregular oceans, it breaks down into a series of tidal systems. Each system consists of a central area of no tide, called the 'amphidromic point', around which the tide rotates as a wave. The crest of the wave (high tide) and the trough (low tide) radiate out from the point and rotate around it, clockwise in the southern hemisphere, counter-clockwise in the north. The tidal systems range in size from covering entire ocean basins such as the North and South Atlantic, to small systems in the Red, North and Black seas, and surrounding the British Isles.

Australian tidal systems

Australia's tides are derived from five tidal systems:

- the southwest Pacific tidal system, centred southeast of New Zealand
- the Southern Ocean tidal system, centred south of Tasmania
- the Indian Ocean tidal system, located off the southwest tip of the continent
- two smaller tidal systems in the Arafura Sea.

The southwest Pacific tidal system affects the entire east coast of Australia. It converges with the Tasmanian tidal system in Bass Strait, which also propagates across the southern coast. The Indian Ocean system produces a tide that travels up the Western Australian coast then slows as it progresses across the shallow northern coast and continental shelf, where it combines with the Arafura systems to deliver tides across the north.

The waves generated by tidal systems are hundreds of kilometres from crest to crest, so they are long, relative to the shallow oceans (4 to 5 kilometres deep). As a consequence, they behave like shallow water waves whose speed and direction are influenced by the depth and shape of the ocean basins and sea floor, particularly when crossing the shallow continental shelves (<150 metres deep). As the waves reach the shallower edges of the continents, they slow down because their speed is inversely related to the water depth: the shallower the water, the slower the speed. At the same time, they are amplified in height because of shoaling, particularly over wide, shallow continental shelves.

Tidal variations

Most tides, which average only 0.2 metre in height in the deep ocean, can range up to several metres high at the shore, several times their height in the deep ocean. Two additional processes can also amplify the tide locally. First, convergence of two tidal systems leads to amplification in the tide. This happens in Torres Strait, Bass Strait and in Broad Sound in central Queensland. Second, in a large embayment, the incoming tide often interacts with the outgoing tide to resonate, and this can cause the tide to reach an extreme height.

Figure 2.13 shows the variation in tidal range around the Australian coast. Tides tend to arrive simultaneously along the southern half of the continent and are all less than 2 metres in range, apart from those in shallow Bass Strait, Westernport and the South Australian gulfs, where they can reach 2 to 4 metres. Across northern Australia, the wide, shallow continental shelf slows progression of the tide as it moves from west to east. Shoaling across the shallow

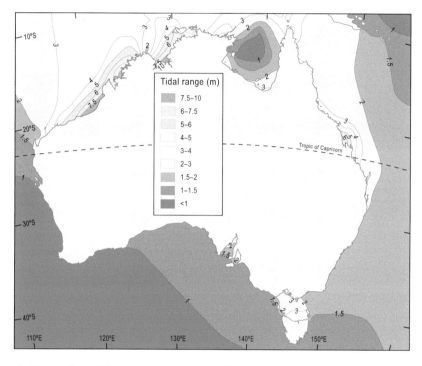

shelf, together with some convergence and resonance, results in all tides exceeding 2 metres, and reaching several metres in some places, particularly in Broad Sound and the northwest coast–Kimberley region. A summary of Australia's tides follows.

Localised tidal systems

In Queensland, the eastern Arafura tidal system rotates clockwise around the Gulf of Carpentaria, delivering generally low tides to the gulf shore, with those in the southeast being predominantly diurnal – that is, experiencing only one high and one low tide a day. Interaction between tides in the Gulf and those in the Coral Sea occurs through Torres Strait, with the tide moving from east to west and being amplified to more than 2 metres.

Figure 2.13 *The range of spring tides around the Australian coast. Note the generally high tides across northern Australia, while tides less than 2 metres dominate across southern Australia, apart from the shallow Bass Strait and the South Australian gulfs.* Source: National Tidal Centre

Particularly strong tidal currents occur through Torres Strait, with velocities well in excess of 2 metres per second, driven by the tidal systems on either side of Cape York. Along the east Queensland coast the tide travels from north to south, increasing in height to the north and slowing as it enters the

Figure 2.14 *Roebuck Bay in the north of Western Australia has an 8-metre spring tide range. In this view, the leading edge of the 300-metre-wide band of mangroves in the centre marks the mean sea level (mid-tide). Spring high tide extends up to 3 kilometres across white salt flats to the rear, and low tide extends across mud flats (lower) more than 3 kilometres into the bay, providing a spring intertidal width of up to 6 kilometres.* Photo: A.D. Short

Broad Sound region, where the convergence of two tides causes amplification up to 8 metres.

Along the New South Wales coast, the tide approaches and crosses the steep, narrow continental shelf and arrives at a similar time and height (spring range 1.3 metres) along the entire coast. In contrast, in Victoria and Tasmania, two tidal systems converge, moving slowly into Bass Strait from both sides, and clockwise around the Tasmanian coast. Tides remain low on the open coast of both states, but are amplified on both sides of Bass Strait to reach 2 to 3 metres. Across South Australia and the south coast of Western Australia, the Southern Ocean tide approaches parallel to the coast and is only slightly amplified across the wide but deep continental shelf,

with some of the world's lowest continental tides (<0.25 metre) along the southern tip of Western Australia. In the South Australian gulfs, the tides are slowed by the shallow seabed, which takes 7 hours to reach Port Augusta and in the process is amplified to 3 metres in upper St Vincent Gulf and 3.5 metres in upper Spencer Gulf.

The Indian Ocean tidal system reaches the southwest coast of Western Australian with a low amplitude (<1 metre). It travels up the coast, and once past Exmouth Gulf, begins to be amplified across the wide, shallow northwest continental shelf, reaching as high as 7 metres along Eighty Mile beach. The tide slows as it moves around the Kimberley coast, where it begins to merge with the western Arafura tidal systems. Australia's highest tides occur at Derby in King Sound (12 metres) due to resonance, at Broome (9 metres) and at Wyndham in Cambridge Gulf (8 metres). The tide progresses from west to east across the Northern Territory coast, with heights between 3 to 6 metres. In some tapering north-coast rivers, the flooding tide becomes highly dominant and amplified, and can produce a tidal bore (see box 'Tidal asymmetry and tidal bores', page 92).

The increasing tidal range causes the position of the water surface to move up and down, sometimes displacing the shoreline hundreds of metres at each tide (see Figure 2.14). It also generates strong tidal currents to move the water to and from the coast. These currents tend to flow north, with the flooding tides along the northwest and northeast coast contributing to the overall northerly sediment transport, particularly on eastern Cape York Peninsula.

Tsunami

A tsunami is a wave, or series of waves, generated by a sudden movement in the ocean's surface, usually caused by an earthquake, volcanic eruption or undersea landslide, or externally by a meteorite impact. Once generated, a tsunami can travel at up to 950 kilometres per hour (the speed of a jet airliner), spreading out in all directions to travel across entire ocean basins in a matter of hours. In the deep ocean, tsunami tend to be low, only a few centimetres high, but 200 to 300 kilometres long, and consequently they contain huge volumes of water. However, as they approach the shallow continental shelf they can be both focused by depth of the seafloor and amplified as the speeding crest begins to catch up with the slowing trough. As a result, a tsunami may arrive at the shore as a high (up to several metres), turbulent wall of water travelling at up to 50 kilometres per hour. This enables it to rush inland for several minutes until it reaches land the same height as its crest, and run up to even higher elevations on steep slopes.

A tsunami event usually consists of several waves, about 15 minutes apart, with each wave flowing inland for several minutes before retreating for several minutes and carrying debris back out to sea. The highest wave may be the first, second or third wave. In many cases, the trough arrives first, causing the waterline to fall, often exposing hundreds of metres of seabed. Unfortunately, this tends to attract people to the bare seabed; while wandering out they encounter the speeding crest of the first incoming wave.

Tsunami are generated in the seismically active Indonesian islands and the north coast of

New Guinea, and in the Pacific in the volcanically active chain of islands including the Solomon Islands, Vanuatu and New Zealand. Earthquakes and volcanic eruptions are common in these areas and usually generate tsunami, as in the case of the tragic Asian tsunami, which centred on Aceh in northern Sumatra on Boxing Day 2004 (see Figure 2.15). Australia also receives highly attenuated tsunami from these regions as well as sources as distant as Chile and Alaska.

Australia's tsunami risk

Fortunately, these major tsunami have significantly less impact on the Australian coast than in their source regions. While Australia records a tsunami on average every other year, with historical records of 47 known events since 1858, most are very small and present only a low risk to life or property. The majority of these tsunami events were recorded on the New South Wales coast, although the most tsunami-prone area of Australia is the northwest coast, facing towards Indonesia. Following the giant Krakatoa volcanic eruption in 1883, a tsunami was generated; it affected most of the Western Australian coast and reached 6 metres in height along a small section of the northwest Australian coast. However, there is some evidence to suggest that, in the past few thousand years, tsunami have reached kilometres inland along the northwest coast.

Overall, Australia is a low-risk tsunami environment. In order to provide warnings of tsunami events, Australia is developing an independent tsunami warning system, supports the Pacific Tsunami Warning System and will actively participate in the Indian Ocean Tsunami Warning System. While these

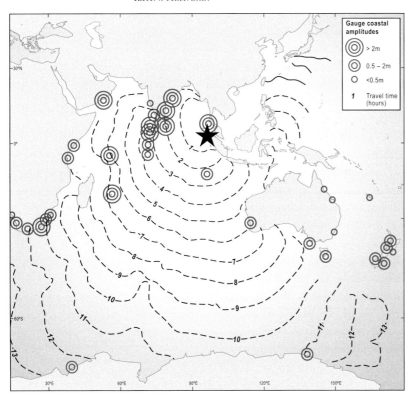

Figure 2.15 *The Boxing Day 2004 tsunami originated off Aceh in northern Sumatra (star) and travelled outwards at approximately 800 kilometres per hour. This map shows the travel times – it reached northern Australia 5 hours later, Adelaide after 10 hours and Sydney about 20 hours later. The waves travelled around the entire planet. The height of the tsunami in the deep ocean was greatest in the northern Indian Ocean, as shown by the red circles. The height diminished to the south and was less than 50 centimetres high when it reached eastern Australia.*

systems cannot stop tsunami, they can, however, detect the seismic activity that generates the tsunami and, using ocean buoys, can determine the height, speed and direction of the tsunami once generated. In this way coastal communities can be forewarned about the approach, time of arrival and likely height of a tsunami. This greater awareness of the threat posed by tsunami has led to evacuation of east-coast beaches in recent years when a tsunami-generating earthquake was detected in the Solomon Islands.

Australian ocean currents

Ocean currents refer to the wind-driven movement of the ocean surface. The ocean surface is that thin layer of surface water that is warmed by solar radiation to a depth of about 200 metres, below which are the cooler, deep ocean waters, that extend to 4–5 kilometre depths. While ocean currents only occasionally have a direct impact on parts of the coast, they do play a major role in coastal climate and ocean temperature.

Australia is located between the warm, easterly flowing South Equatorial Current, which is driven across the equatorial Pacific Ocean by the easterly trade winds, and the westerly flowing, cool Antarctic Circumpolar Current in the Southern Ocean, driven by the strong westerly winds (see Figure 2.16). The northern part of the continent intercepts the equatorial current and deflects the bulk of the warm current south along the east coast. This is the East Australian Current, a warm current (20–24° Celsius) that also receives an input of warm water from the Great Barrier Reef lagoon. The East Australian Current flows as a continuous current at

speeds up to 7 kilometres per hour along the edge of the continental shelf, south to Seal Rocks in New South Wales, where it then breaks into a series of counter-clockwise-rotating, warm core eddies up to 100 kilometres in diameter. These eddies spiral south along the coast and out into the Tasman Sea, forming the Tasman Front, which can bring warmer water to Lord Howe Island. The core of the current, however, continues south to Tasmania.

In the north, part of the South Equatorial Current continues as a constricted and weaker current flowing through Torres Strait and across the Arafura Sea. This then links with the Indonesian Throughflow in the Timor Sea, which it joins to reconstitute the equatorial current south of Indonesia. These warm tropical currents have an average temperature higher than 25 degrees Celsius, commonly reaching 30 degrees Celsius during summer. Part of the current tracks along the northwest coast and, under the influence of the Coriolis effect, hugs the Western Australian coast down to Cape Leeuwin. This current, known as the Leeuwin Current, flows south at up to 6.5 kilometres per hour and brings warm water (20–26 degrees Celsius) down the coast. It is responsible for the most poleward coral reef systems in the Indian Ocean, on the coast at Ningaloo (24 degrees South) and offshore in the Houtman Abrolhos archipelago (29 degrees South). As the Leeuwin Current rounds Cape Leeuwin it usually breaks into a series of warm core, counter-clockwise eddies that move across the south of the continent. They can even be tracked in patches right across the south coast to Bass Strait, where the still warmer water flows down the continental slopes into the Tasman Sea.

Across the south, the great Antarctic Circumpolar Current is a broad, deep current that

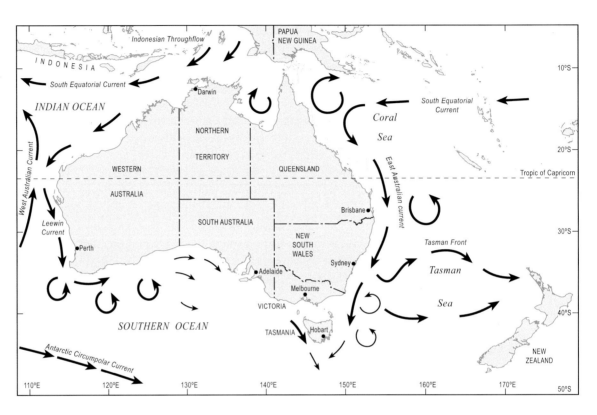

Figure 2.16 *The major surface ocean currents in the Australian region. Global wind systems drive the major equatorial and circumpolar currents, with the continent deflecting the East Australian Current. Warm water and the Coriolis effect drives the Leeuwin Current, while local and seasonal winds drive some coastal currents. The continental shelf is shaded white.*

moves cool water (4–12 degrees Celsius) continuously around Antarctica. It flows across the Southern Ocean well south of Australia, with no direct impact on the coast. The Australian coast is therefore dominated by tropical waters, or waters of tropical origins in the north, west and east, with only the southern coast influenced by cooler sub-polar waters.

Sea-surface temperature

The temperature of the sea surface around Australia depends upon the latitude, season and ocean currents. Surface waters are warm in the tropical north and become cooler as latitude increases. They are also seasonally cooler in winter (see Figure 2.17). During summer, warm water (>28 degrees Celsius) extends across the north coast between Broome and Cairns, with temperature decreasing southward to 20 degrees Celsius at Cape Leeuwin in Western Australia and Cape Howe in Victoria, and averaging 18–20 degrees Celsius across the south coast. In winter, temperatures drop to 22–24 degrees Celsius across the north, 16 degrees Celsius at Cape Leeuwin and Cape Howe and 12–16 degrees Celsius across the south. The Leeuwin and East Australian current have an impact on both coasts, bringing warmer water down the coasts off the south of Western Australia and along the New South Wales coast (see Figure 2.17).

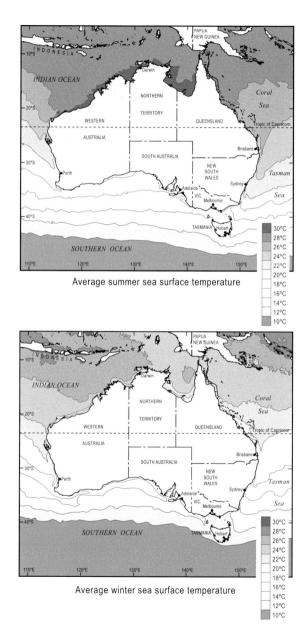

Average summer sea surface temperature

Average winter sea surface temperature

Figure 2.17 *Summer and winter sea surface temperatures in the Australian region. Note the latitudinal zonation in temperature, its seasonal range and the movement of warmer tropical waters down both the east and west coasts.*

Conclusion

The processes described in this chapter are part of Australia's atmospheric and marine climates, and at the coast they combine to provide the energy that shapes our coast and continually changes the shoreline. The atmospheric climate provides the rainfall and temperature that influence the denudation of the entire continent and supply of sediment to the coast, as well as all associated ecosystems. The wind drives the ocean currents and generates the waves that pound the coast, and at the shore blows sand inland to form dunes. Around the coast, ocean temperature, waves, tides, ocean currents and wind all contribute energy to form and shape the coast. They set the habitat boundaries within which the coastal ecosystems flourish.

Whereas Australia has an arid core, its coastline has a wide variety of coastal systems, ranging from those dominated by high waves and low tides in the temperate south, to the northern tropical systems, where low waves prevail and large tides dominate. The result is a wide range of coastal ecosystems including dune vegetation, salt marshes, mangroves, seagrasses and coral reef systems, which are described in the next chapter. The subsequent chapters examine the major physical systems and their distribution and variation around the coast, namely estuaries, beaches, dunes, rocky coasts and coral reefs.

COASTAL ECOSYSTEMS

AUSTRALIA HAS A wide range of coastal ecosystems including coral reefs, mangroves, seagrass meadows and coastal dune vegetation. All are influenced by both the atmospheric and marine climates and each consists of a wide range of interdependent flora and fauna that requires a certain level of nutrients, and is linked to adjacent food chains. Each ecosystem can tolerate a certain range of environmental factors. On land, air temperature and precipitation are critical, together with humidity and wind, and the nature of the land surface. At the shoreline, wave energy, tidal range, water temperature and substrate become critical, while in the subtidal, these factors plus water depth exert control.

Australia's atmospheric and marine climates therefore control the nature and range of its coastal ecosystems. The systems have also evolved on a remote southern continent, surrounded by oceans and seas, and while many marine species can disperse widely, the net result is a distinctive Australian coastal ecology that varies around the coast. In southern Australia, there is a significant level of uniqueness, called endemism, in the coastal ecology. This endemism reflects the isolation of southern Australian ecosystems, whereas northern Australia shares ecosystems with the Indo-Pacific region.

The Australian coast is surrounded by rich terrestrial and marine ecosystems, which range from tropical to temperate. On higher-energy, open coasts these habitats include coastal dunes, the beaches and adjacent rocky shores. In more sheltered areas we find that coastal wetlands, salt marshes, mangroves and seagrasses are prolific, and in the tropics there are very extensive coral reef systems. All of these environments occupy an often narrow zone between the deeper ocean and the backshore, and are impacted variously by breaking waves, tidal oscillation and currents, as well as by the regional atmospheric and marine climates.

As a consequence, there is considerable variation within and between all systems. For example, in some places, coastal dunes grade from exposed hardy grasses on their seaward edge to tropical rainforests in the hind dune area. In other areas, sheltered intertidal areas grade from low salt marshes to dense forests of tall mangroves across northern Australia.

Likewise, the ecologies of beaches, rocky shores and shallow seabeds are patterned according to location, elevation/depth and climate. In this chapter we provide an overview of each of these ecosystems, starting on land and moving shoreward and seaward, and focusing on the dominant living organisms and their main regional variations around Australia.

Coast dune vegetation

Coastal sand dunes are inhospitable environments, for several reasons: they are composed of porous infertile sand; they are exposed to strong winds and salt spray; plants can be drenched in seawater and buried by sand; and periodically, the front of the dunes can be eroded by waves. As a consequence only a relatively few, highly specialised plants are able to grow on dunes. Those that do, however, have adapted extremely well and thrive in this harsh environment.

In adapting to this environment, dune vegetation varies at two scales: transitional and regional adaptation.

Transitional adaptation

There is always a predictable distribution by zone or transition in the type of dune vegetation, from the exposed beach to the more sheltered inland (see Figure 3.1). This begins at the beach, where no plants grow in the active wet zone that is regularly swept by waves. Immediately behind the limit of wave impact, hardy grasses, succulents and some creepers begin to take hold. These plants always have a low profile and can tolerate strong winds, salt spray, saltwater immersion and burial by sand. In fact, they thrive as a consequence of sand burial, growing upwards through the fresh layers of sand, and leaving below a network of roots that helps to bind the sand and supply nutrients. This first line of plants is called the 'primary species'. The zone they occupy is called the 'incipient foredune' – the foremost part of the dune that is the youngest and is expected to be eroded by waves every few years.

The primary plant species are in turn backed by secondary species, composed mainly of low shrubs, which occupy the established foredune. On the exposed seaward face, the shrubs are still subject to strong winds and occasional sand burial; they respond by adopting a low or stunted growth form, usually less than a metre in height. On the leeward side of the foredune, where they are more sheltered from wind and spray, the same shrubs grow erect and taller, perhaps reaching a few metres in height. The vegetation canopy across the foredune often shows a gradual increase in height, providing clear evidence of the importance of wind in shaping plant growth.

The area behind the foredune is called the 'hind dune', and it may take a range of forms. However, in terms of vegetation it is occupied by tertiary, or climax, vegetation for that region. In humid tropics and temperate environments this might be a forest or woodland with tall trees, whereas in arid regions, shrubs, usually less than 2 metres tall, or occasional mallee, comprise the climax vegetation.

Figure 3.1 At The Granites in South Australia, the beach dune displays the typical gradation from an exposed beach to a line of seaweed wrack, to the Spinifex-covered, low incipient foredune and the shrub-covered front of the higher foredune. Photo: A.D. Short

Regional adaptation

The second scale of variation in dune vegetation is regional, with the actual species dependent on climate and biogeography. Although the growth forms and structure remain similar around the Australian coast (that is, grasses, shrubs, trees), the species vary considerably, from north to south and east to west. In the following section we provide a general overview of some of the dominant regional species.

Zonation in the east

Along the east coast there is a transition from tropical Queensland to more temperate dune species in New South Wales, Victoria and Tasmania. The typical dune zonation on the coast, and the change in species from the more tropical north to the more temperate south of New South Wales (see Figure 3.2), are as follows.

1. In moving from the beach inland, the primary species located immediately adjacent to the bare beach sand are predominantly the Beach Spinifex grass (*Spinifex sericeus*) and the exotic, seasonal succulent Sea Rocket (*Cakile maritima*) (see Figure 3.3).

2. These are backed on the foredune by the shrubs Coastal Wattle (*Acacia longifolia*), Teatree (*Leptospermum laevigatum*) and Coastal Banksia (*Banksia integrifolia*).

3. Behind these, in the more sheltered hind dune area, are the eucalyptus species, Pink Bloodwood (*Eucalyptus intermedia*) and Moreton Bay Ash (*Eucalyptus tessellaris*) to the north, and to the south Bangalay (*Eucalyptus botryoides*), Blackbutt (*Eucalyptus pilularis*) and Smooth Barked Apple (*Angophora costata*), together with the tall Old Man Banksia (*Banksia serrata*) and a ground cover of Bracken fern (*Pteridium esculentum*).

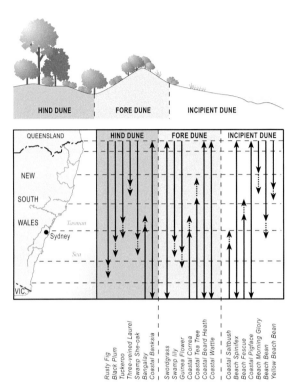

Figure 3.2 *Dominant species associated with coastal dunes in New South Wales display two spatial trends. Moving inland from the bare sand beach, grasses and succulents dominate the incipient foredune, while shrubs dominate the foredune, and shrubs grade into trees at the hind dune. Second, moving from north to south, there is a gradation from more tropical to temperate species in all three zones.*

Figure 3.3 *Dune vegetation at Seven Mile Beach on the southern New South Wales coast, with Spinifex grass covering the incipient foredune, then a zone of low Acacia shrub, backed by taller Teatrees.* Photo: A.D. Short

Figure 3.2 also illustrates the change in species in each of the three zones, when one moves from north to south along the coast:

1. On the incipient foredune, while Beach Spinifex thrives along the entire coast, Beach Morning Glory (*Ipomoea pes-caprae*), Pink Bean (*Canavalia maritima*) and *Vigna* are restricted to the north of the state, and Grey Saltbush (*Atriplex cinerea*), Beach Festuca (*Austrofestuca littoralis*) and the introduced Marram grass (*Ammophila arenaria*) to the south.

2. On the foredune, some plants also grow the length of the coast while others like the trailer Guinea Flower (*Hibbertia scandens*) are restricted to the north and Teatree to the south coast, with most species overlapping along the central coast.

3. In the hind dune, only the Bottle Brush shrub grows the length of the coast, with the Bangalay restricted to the south, and the remainder increasing to the north.

The typical species of southern New South Wales continue into eastern Victoria and much of Tasmania, while the introduced European Marram grass increases at the expense of the local Spinifex.

Semi-arid and arid coastal vegetation

Coastal dunes along semi-arid to arid South Australia are vegetated by a sequence of hardy plants, beginning with herbs and grasses on the foredunes and grading landward into a combination of sedgelands and shrublands, with trees limited to a few locations. The typical sequence is as follows:

1. At the incipient foredune and outer foredune are the annual succulent Sea Rocket (*Cakile maritima*), backed by the prolific Beach Spinifex (*Spinifex sericeus*), which is often accompanied by the exotic spurge *Euphorbia paralias*, particularly in the east.
2. The main foredunes and hind dunes are dominated by a range of generally low (<2 metres) shrublands, with the more important including Grey Saltbush (*Atriplex cinerea*), Bladder Saltbush, (*Atriplex vesicaria*), Marsh Saltbush (*Atriplex paludosa*), Cushion Bush (*Leucophyta brownii*), Black Tea-tree (*Melaleuca lanceolata*), Brown Spinach (*Tetragonia implexicoma*), Nitre Bush (*Nitraria billardierei*) and Coastal Daisy occurring along many of the gulf and western dune fields.
3. Mallee trees such as the Soap Mallee (*Eucalyptus diversifolia*) only grow in some more protected hind-dune locations.

Coastal dunes occur around the entire Western Australian coast. In the southwest, the beach spinifex community – Hairy Spinifex (*Spinifex hirsutus*),

Spinifex longifolia, Sea Rocket (*Cakile maritima*) and Beach Daisy (*Arctotheca populifolia*) – dominates the incipient foredune. The seaward side of the foredune usually has low acacia shrubs (*Acacia cyclops*, *Acacia rostellifera* and Coastal Daisy), with larger shrubs on the crest and leeward side, which grade into a stable shrub community (*Acacia rostellifera*, Coastal Daisy). Only in the humid southwest do Tuart forests (*Eucalyptus gomphocephala*) form the climax dune vegetation.

In the Kimberley region, the incipient foredune is dominated by the Beach Spinifex community and sedges (*Fimbristylis cymosa*, *Fimbristylis sericea*, *Cyperus bulbosus*). The foredune has low shrubs (*Acacia bivenosa*, *Lysiphyllum cunninghamii*, Beach Bean (*Canavalia rosea*)), while the hind dune and hollows have a dense shrub community of diverse plants, including the ubiquitous Coastal Screw Pine (*Pandanus spiralis*).

Without a doubt, the considerable diversity in species around the Australian coast is a reflection of the range of coastal climates from tropical to temperate to arid – as well as the geographical isolation of the coasts – that have permitted endemic species to evolve. We look at the role of dune vegetation in the nature and formation of coastal dunes in more detail in Chapter 6.

Freshwater wetlands on coastal plains

Freshwater wetlands occur at the landward margin of many coastal and estuarine plains. The most extensive and best known are those in Kakadu National Park, but similar wetlands also occur along many of the tidal rivers of northern Australia, particularly in the monsoonal Top End and along the lower reaches of rivers and streams in northeastern Queensland. In southern Australia, such wetlands are restricted to waterlogged areas between beach ridges and in saline back-barrier depressions and lagoons. They provide a stark contrast in species, biomass and extent to the northern wetlands.

The low-lying plains of the Kakadu National Park contain productive grass and sedge wetlands. Flooding of the plains occurs seasonally, as a result of direct rainfall and augmented by floodwaters from the rivers. Wild rice (*Oryza rufipogon*) occurs over much of the plains, as do other perennial grasses (such as *Hymenachne acutigluma* and *Pseudoraphis spinescens*). Sedges, such as species of *Eleocharis*, *Fimbristylis* and *Cyperus*, are also widespread.

The patterning of vegetation in Kakadu is complex, and is related mostly to subtle differences in elevation and, consequently, duration of flooding. Thus, vegetation may vary from year to year. Through the year, the plains dry out incrementally, with the lowest-lying areas, such as billabongs, remaining flooded late into the dry season. Only a few billabongs persist, with their soils remaining waterlogged for much of the year. These deeper wetlands contain waterlilies (species of *Nymphaea* and *Nymphoides*), Lotus (*Nelumbo nucifera*) as well as floating plants (such as *Azolla*). They support fish, freshwater turtles and saltwater and freshwater crocodiles, as well as impressive numbers of waterbirds, including large numbers of magpie geese, ducks and waders such as jabiru and brolga.

Freshwater wetlands are also found on the hinterland margin of coastal plains elsewhere in eastern Australia. In northeast Queensland, there are similar wetlands, although with the Paperbark (*Melaleuca quinquenervia*) dominating the woodlands. The reed *Phragmites* is commonly found in more temperate areas, and forests of Sheoak (*Allocasuarina glauca*) are typically found landward of the salt marsh and upstream of the tidal limit.

Salt marsh and samphire vegetation

Salt marshes grow on sheltered areas of waterlogged intertidal sediments as salt-tolerant grasses, reeds and small shrubs. They lie between neap and spring high tide, and although they are regularly covered by spring tides, they can also extend landward into the supratidal zone, which is that area of the shore that is only reached by the tide a few days each month, if at all. Salt marshes can tolerate the occasional salt intrusion, but prefer lower salinities. They occur worldwide and cover an area of 13 500 square kilometres around Australia (see Figure 3.4). They are most extensive – but low in species diversity – in the tropics. With increasing latitude they begin to multiply in richness of species and complexity of community, resulting in the best-developed systems in the temperate south. More than 200 species have been identified around Australia, although at any one location usually only 4–6 species occur. The vegetation is dominated by plants from four families: grasses (Poaceae), saltbushes (Chenopodiaceae), rushes (Juncaceae) and sedges (Cyperaceae).

In northern Australia, saltmarshes form a mosaic of succulents, grasses, low shrubs and samphire that occur to landward of often more extensive mangrove forests or adjacent to salt flat

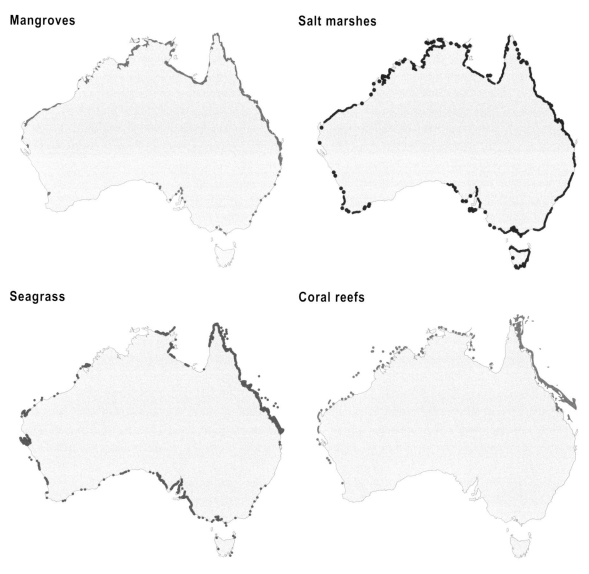

Figure 3.4 **a)** *Distribution of mangroves around Australia. The number of species increases from none and one in the south to 40 in the north. Size, biomass and area also increase to the north;* **b)** *Distribution of Australian saltmarsh species. Note the dominance in the more temperate latitudes of southern Australia. The area covered by saltmarshes is, however, more extensive in northern Australia;* **c)** *Distribution of tropical and temperate seagrass species around Australia;* **d)** *Distribution of coral reefs across northern and western Australia.*

areas. These are extreme, inhospitable environments with highly variable salinities, and no shade or protection from wind. Particularly conspicuous plants are the succulents Beaded Glasswort (*Sarcocornia quinqueflora*), Saltwort (*Batis argillicola*), Sea purslane (*Sesuvium portulacastrum*) and other samphires (*Tecticornia australasica*, *Suaeda* spp. and *Halosarcia* spp.). In New South Wales, grasses tend to dominate the salt marshes, whereas across Victoria, Tasmania, South Australia and southern Western Australia shrubby plants are more dominant.

Along more arid sections of southern and western Australia, samphire vegetation grows in damp, waterlogged supratidal coastal areas in association with algal mats, particularly in the lower swales between beach ridges and in the saline back-barrier depressions and dry lagoons. Samphire vegetation comprises those hardy plants that live in the supratidal zone, usually grasses, succulents and low shrubs (<1 metre). Samphire vegetation usually forms a boundary between the shoreline and the landward terrestrial vegetation. In South Australia, the samphire vegetation is dominated by the saltbush family, Chenopodiaceae. In the southwest of Western Australia, there is a transition from high water, landward of the following three communities: the low, succulent Chicken Claws (*Salicornia australis*) extends close to high water; this is backed by the Grey Samphire (*Tectacornia halocnemioides*) community, a low woody, succulent glasswort shrub, in the zone rarely reached by high tide. The sedge zone of tussocky grasses occupies the most landward position, and is dominated by Coast Saw sedge (*Gahnia trifida*), Searush (*Juncus kraussii*) and knobby clubrush (*Isolepis nodosus*).

In the Pilbara and Kimberley regions, tides are higher, waves lower and the climate hotter, resulting in wide intertidal to supratidal zones and greater climate stress. Four communities of samphire plants can occur in favourable locations. At the seaward fringe a community of succulent samphires, then a mixed community of herbs and grasses in the mid to upper-marsh level. Herbs and low shrubs occupy the higher, well-drained locations, while the most landward community, which can tolerate the high salinity but not waterlogging, include Desert Grasswort (*Neobassia astrocarpa*), Coastal Spinach (*Trianthema turgidifolia*) and the Bunch Grass (*Triodia* sp.).

Mangroves

Mangroves are a diverse range of trees, shrubs and palms that grow between mean sea level and the level of the highest tide (see Figure 3.5). They have all adapted to survive in this otherwise inhospitable, salty environment. Mangroves are particularly extensive on sheltered, macrotidal tropical shorelines, but extend into subtropical areas, and occur in some isolated estuaries along the southern coast of Australia, but not in Tasmania – in total occupying about 18 per cent of the Australian coast (see Figure 3.4 and Table 3.1). The term 'mangrove' is applied to the individual tree as well as to the forest as a whole, which is sometimes also referred to as 'mangal'.

Mangroves, like saltmarsh vegetation, are halophytes, which means they can survive in salty conditions through a range of adaptations. Some mangroves exclude salt from their roots, whereas others expel salt, which then accumulates on the leaves. Their root systems have also adapted to the

Figure 3.5 *A 200-metre-wide band of mangroves in the southern Gulf of Carpentaria, located between mean sea level and the highest tide level. They are fronted by wide mud flats and backed by salt flats extending inland across the supratidal zone.* Photo: A.D. Short

Figure 3.6 *The Stilt Mangrove* Rhizophora *forms a prominent zone in the mangroves of tropical Australia. The forest, as in this stand of* Rhizophora stylosa *on the southern shore of van Diemen Gulf, forms an almost impenetrable thicket, and the muddy substrate is flooded during most high tides.* Photo: C.D. Woodroffe

waterlogged substrates, with some prop roots (see Figure 3.6) providing support for the tree as well as being covered with cells through which the plant takes in air. The widespread genus *Avicennia*, the Grey Mangrove, has pencil-like breathing roots that rise up out of the mud, called 'pneumatophores', and the Apple Mangrove *Sonneratia* has similar but larger roots. The Stilt Mangrove *Rhizophora* has arching prop roots (see Figure 3.6), the Orange Mangrove (*Bruguiera*) has knee roots and the Yellow Mangrove (*Ceriops*) is characterised by buttresses or smaller knee roots. Most species have viviparous propagules, which means that their seeds remain attached and germinate on the tree, and are buoyant during a short period of dispersal.

Location and diversity of mangroves

Mangroves are found around most of the Australian mainland, apart from a 2500-kilometre section from Davenport Creek in South Australia to Bunbury in Western Australia, and Tasmania. Broadly, their distribution is controlled by temperature, with mean air temperatures of the coldest month generally >20 degrees Celsius, while their intolerance of winter frosts sets the latitudinal limit. Species diversity and height also decrease with latitude. Only one genus, *Avicennia* (Grey Mangrove), occurs along the south coast, located in isolated stands around Bunbury in Western Australia, in the bays and gulfs in South Australia and at the southernmost stand of mangroves in the world: at 38 degrees 45 minutes South, at Corner Inlet in Victoria, where it occurs as a low (<1 metre) sprawling shrub.

The greatest number of mangrove species (34) occurs in north Queensland, with species numbers decreasing to the west and to the south (see Table 3.2) of the continent. The most prolific mangrove forests, both in terms of area and species diversity, are found in the monsoonal tropical northern coast. Of Australia's 11 560 square kilometres of mangroves, 88 per cent is spread relatively evenly between the Kimberley, Northern Territory and Queensland. *Avicennia* grows 20–30 metres tall in places in northern Australia, decreasing to less than a metre in Victoria. The wetter east coast has more diverse mangrove forests than the drier west coast at equivalent latitudes, where they are limited by hypersaline groundwaters.

Northern mangrove communities

The tropical climate of northern Australia permits a greater number of mangrove species, and also results in taller mangroves (10–30 metres) with a greater

	km²	%
Queensland	4602	39.9
New South Wales	99	0.8
Victoria	12	0.1
Tasmania	0	0.0
South Australia	211	1.9
Western Australia	2517	21.7
Northern Territory	4119	35.6
Australia	11 560	100.0

Table 3.1 Mangrove area in Australia

biomass. Mangrove communities in the north tend to be diverse, tall and relatively wide, the latter assisted by the high tidal ranges leading to wide intertidal areas. As a consequence of these factors, the area of mangroves, number of species and biomass increase substantially from south to north. Australia-wide, 96 per cent of mangroves are located in the tropics north of the Tropic of Capricorn. Western Australia has more than 2500 square kilometres of mangroves along the mainland, with another 90 square kilometres on islands, which in total comprise more than 20 per cent of Australia's mangroves, most located on the Kimberley coast (see Table 3.1). At the global level, Australia has the world's third-largest area of mangroves, after Brazil and Indonesia.

While mangroves occur on sheltered but open-ocean shorelines in northern Australia, the different species grow at different elevations within the tidal range, and the mangrove forest appears zoned. The zonation shown in Figure 3.7 is typical, with seaward woodlands of *Sonneratia alba* and isolated individuals of *Avicennia marina* rooted close to mean sea level. There is generally a tall forest of *Rhizophora* landward of this, made impenetrable by a tangle of prop and aerial roots, and subject to flooding by the majority of tides (see Figure 3.6). In the upper intertidal zone, *Rhizophora* is replaced by *Bruguiera*, and then by a thicket of *Ceriops*, sometimes with a bare hypersaline flat in its midst. At the landward margin, *Avicennia* is once again prominent, together with salt marsh plants.

The value of mangroves

In the past, there was a perception that mangrove forests were wastelands, and mangroves were often cleared and the land used for agriculture, industry

Figure 3.7 *Mangrove zonation, as indicated by the change in colour, Cape Bowling Green, Queensland. The light-coloured canopy along the shoreline is dominated by the Apple Mangrove (*Sonneratia*) and the Grey Mangrove (*Avicennia*), whereas the middle zone contains* Rhizophora, *with a zone in which many of the trees are dead, replaced to landward by less salt-tolerant species.* Photo: A.D. Short

Table 3.2 Distribution of mangroves species around Australia

Species	SW Western Australia	Central Western Australia	Pilbara	Kimberley	Bonaparte	Top End	West Gulf	East Gulf	NE Queensland	Central Queensland	SE Queensland	Central New South Wales	New South Wales	South New South Wales	Victoria	South Australia
Acanthus ebracteatus				x	x	x	x	x	x	x						
Acanthus ilicifolius					x	x	x	x	x	x						
Acrostichum speciosum			x	x	X	x	x	x	x	x	x	x				
Aegialitis annulata			x	x	x	x	x	x	x	x	x					
Aegiceras corniculatum			x	x	x	x	x	x	x	x	x	x	x	x		
Avicennia integra						x										
Avicennia marina	x	x	x	x	x	x	x	x	x	x	x	x	x	x	x	x
Bruguiera cylindrica									x	x						
Bruguiera exaristata									x	x						
Bruguiera gymnorrhiza			x	x	x	x	x	x	x	x						
Bruguiera parviflora					x	x	x	x	x	x	x	x				
Bruguiera sexangula				x	x	x	x	x	x	x						
Camptostemon schultzii			x	x	x	x	x	x	x							
Ceriops australis			x	x	x	x	x	x	x	x	x					
Ceriops decandra						x		x	x							
Ceriops tagal						x		x	x	x						
Cynometra iripa						x	x	x	x	x						
Diospyros littorea				x	x	x	x	x	x							
Dolichandrone spathacea									x	x						
Excoecaria agallocha				x	x	x	x	x	x	x	x	x	x			
Heritiera littoralis								x	x	x						
Lumnitzera littorea							x	x	x	x						
Lumnitzera racemosa				x	x	x	x	x	x	x	x					
Nypa fruticans						x			x	x						
Osbornia octodonta			x	x	x	x	x	x	x	x	x	x				
Rhizophora apiculata						x		x	x	x						
Rhizophora lamarckii						x		x	x							
Rhizophora mucronata								x	x							
Rhizophora stylosa		x	x	x	x	x	x	x	x	x	x	x				
Scyphiphora hydrophylacea				x		x	x	x	x	x	x					
Sonneratia alba					x	x	x		x	x	x					
Sonneratia caseolaris																
Sonneratia lanceolata									x	x						
Xylocarpus granatum					x		x		x	x	x	x				
Xylocarpus moluccensis					x	x	x	x	x	x	x	x				
Total	1	2	9	18	17	3	20	33	34	21	13	6	3	2	1	1

or even residential development. In the worst cases, the 'swamp' was used as a rubbish tip, particularly in some of the bays around Sydney Harbour. Over recent decades the important ecological roles of mangroves have been realised and mangroves are now protected throughout Australia.

Mangroves are highly productive. They grow at similar rates to other trees, and their timber, leaves and fruit support a range of ecosystems. There are many fish species that spend early stages of their life cycle in and around mangrove forests, and healthy intertidal wetlands are important for the barramundi and shrimp fisheries of northern Australia. Mangroves provide a habitat for crocodiles (see box 'Mangroves and crocodiles', right). Other animals also live in or around the mangroves, including molluscs and crustaceans (such as the mud crab and the mud lobster). Mangrove wood supports ship worms (*Teredo*), and large populations of crabs feed on their leaves, which, when broken down, contribute to the estuarine detrital food web.

In addition to the regular flooding by tide, and the longer-term geomorphological processes of erosion and sediment deposition, mangrove forests are subject to the impacts of extreme events, particularly tropical cyclones. Such storms can uproot mangroves, leaving thickets of *Ceriops* prostrate, resembling matchsticks lined up in the direction of the wind that felled them. Other mangrove species may be defoliated and killed, but remain standing. The generation of gaps in the forest, whether by storm or by lightning strike, may set off a pattern of regeneration that gives rise to the mosaic of mangrove species. Or, as after Cyclone Tracy that hit Darwin on Christmas Day in 1974, these gaps may result in broad clearings or swathes of felled trees that persist for decades.

MANGROVES AND CROCODILES

The tidal creeks and mangrove forests of northern Australia are a favoured habitat of the saltwater crocodile (*Crocodylus porosus*). Crocodiles have been protected in Australia since 1970, and it is now estimated there are 20 000 saltwater crocodiles in the Kimberley, 75 000 in the Northern Territory and 20 000 in Queensland, representing about half the world's crocodile population. Saltwater crocodiles are aggressive and must be avoided, particularly in the water. While they tend to live in tropical tidal creeks and rivers, they are also commonly found on beaches, at sea and on coral reefs far from shore. So beware.

Saltwater crocodiles average 4 metres in length, but can reach 7 metres. They are an opportunistic feeder, usually lying in wait and attacking at night. They have a broad diet of fish, turtles, young crocodiles and sharks, and animals including dogs, kangaroos, wild pigs, dingos, water buffalo, cattle, horses, goannas and birds. They are most active and aggressive in the summer breeding season. Fortunately, being reptiles, they eat only occasionally, sunning themselves on riverbanks for days while they digest their food. They also tend to hibernate in the cooler winter months. Regrettably, on average one person is attacked by crocodiles in northern Australia each year, in most cases where the person ventures into the crocodile's watery domain. The best way to avoid crocodiles is to stay well clear of tidal creeks, river banks and estuarine waters, even many kilometres inland.

Seagrasses

Seagrasses are highly specialised, marine flowering plants that have adapted to living underwater in soft sediments, in shallow nearshore environments (see Figure 3.8). They are a higher order of vascular plants, distinct from seaweeds, which are marine algae. Seagrasses produce flowers and seeds and have rhizomes and roots. They require light for photosynthesis, and do not tolerate high agitation by waves and currents. As a consequence, seagrass meadows thrive in clear, shallow sandy seabeds (usually to a maximum of 10 to 20 metres depth, but in places found in 50 metres water depth in Western Australia), sheltered from ocean wave action.

Australia has the most extensive tropical and temperate seagrass meadows in the world, occupying at least 55 000 square kilometres (see Table 3.3 and Figure 3.4). Their widespread distribution is due to the long east–west coastlines in the tropical north and temperate south, and the extensive sheltered sections of coast containing clear water. Around the southern Australian coast, these conditions are found in large bays (Moreton, Jervis, Phillip, Streaky bays) and numerous estuaries, in St Vincent (1530 square kilometres) and Spencer (3700 square kilometre) gulfs and along parts of the Great Australian Bight and the central Western Australian coast, where they are sheltered by shallow reefs. Australia's seagrass meadows reach their greatest extent (4500 square kilometres) in the clear, shallow waters of Shark Bay. Across northern Australia, the extensive shallow water and low waves provide suitable habitats from the Kimberley around to Hervey Bay, with the most extensive tropical seagrass meadows found in Torres Strait and the Gulf of Carpentaria.

	km²	%
Queensland	22 300	40.1
New South Wales	155	0.3
Victoria	100	0.2
Tasmania	500	0.9
South Australia	9620	17.3
Western Australia	22 000	39.6
Northern Territory	900	1.6
Australia	55 575	100.0

Table 3.3 *Distribution of seagrass around Australia*

The distribution of seagrass species is shown in Table 3.3. The tropical and temperate species are largely restricted to northern and southern Australia, respectively. However, some tropical species do occur in suitable habitats along the central Western Australian coast and in northern New South Wales, where the warm Leeuwin and East Australian ocean currents deliver warmer waters. There are 15 tropical species of seagrass and, while the meadows are highly diverse, tropical seagrasses have a lower biomass than the temperate species. The southern coast has five common genera of seagrass (*Posidonia*, *Heterozostera*, *Zostera*, *Amphibolis* and *Halophila*) and 22 temperate species, with the highest biomass, species diversity and endemism found on the southwest of Western Australia, where 19 species occur (see Table 3.4).

Seagrass meadows support a range of microscopic organisms and algae that grow on the leaves, as well as organisms attached to the grass, and benthic fauna that live in the meadows. They are also a nursery for fish such as luderick, bream and snapper, and crustaceans such as crabs, lobsters and prawns. Seagrass meadows are an important habitat for large fish, and are the

Figure 3.8 *Seagrass debris washed onto the beach west of Eucla in Western Australia, to form a seagrass berm. Similar berms extend along tens of kilometres of the coast along the Great Australian Bight.*
Photo: A.D. Short

	SW WA	Shark Bay	Kimberley– Northern Territory	Gulf of Carpentaria – Torres Strait	NE Queensland	SE Queensland	SE Australia	Southern coast
Tropical species								
Halodule uninervis		x	x	x	x	x		
Halodule pinifolia			x	x	x			
Cymodocea angustata		x		x				
Cymodocea rotundata			x		x			
Cymodocea serrulata		x	x	x	x	x		
Syringodium isoetifolium	x	x	x	x	x	x		
Enhalus acoroides			x	x	x			
Thalassodendron ciliatum			x	x	x			
Thalassia hemprichii			x	x	x			
Halophila ovalis	x	x	x	x	x	x	x	
Halophila ovata		x	x	x	x			
Halophila decipiens	x	x	x	x	x	x	x	
Halophila spinulosa		x	x	x	x	x		
Halophila tricostate			x	x	x			
Temperate species								
Amphibolis griffihii	x							x
Amphilbolis antartica	x	x					x	x
Halophila australis	x						x	x
Hererozostera tasmanica	x						x	x
Hererozostera nigricaulis	x							
Hererozostera polychlamys	x							
Zostera mucronata	x							x
Zostera Muelleri							x	x
Zostera capicorni						x	x	
Posidonia sinuosa	x	x						x
Posidonia coriacea	x	x						x
Posidonia australis	x	x					x	x
Posidonia denhartogii	x							x
Posidonia augustifolia	x	x						x
Posidonia robertsonae	x							
Posidonia ostenfeldi	x							x
Ruppia megacarpa	x						x	x
Thalassodendron pachyrhizum	x							
Total	19	13	13	13	13	7	9	13

Table 3.4 Distribution of major tropical and temperate seagrass species around Australia

SEAGRASS AND DUGONGS

Certain species of seagrasses provide an essential source of food for grazers, particularly in northern Australia, where the grazing herbivores include dugongs (*Dugong dugon*) and green turtles (*Chelonia mydas*). Since Australia has extensive seagrass meadows, it also has the world's largest population of dugongs (estimated at 100 000). Dugongs are associated with seagrass meadows from Shark Bay (~10 000 dugongs) in the west to Moreton Bay in the east, with an estimated ~12 000 in the shallow waters of Torres Strait, 10 000 inhabiting the 4000 square kilometres of seagrass meadows in the northern Great Barrier Reef and ~900 in Moreton Bay adjacent to Brisbane.

Dugongs are seagrass community specialists. While feeding they uproot low biomass seagrasses such as *Halodule* and *Halophila* and modify the seagrass beds. They grow up to around 3 metres in length, weigh up to 400 kilograms and live up to 70 years. As mammals they must surface to breathe, but never land. Dugongs are also called 'seacows', and are probably the origin of the myth of the mermaid.

only food for some turtles and all dugongs (see box 'Seagrass and dugongs', above). Fortunately, all Australian seagrass meadows are protected.

There has been a decline in the area of seagrass in many temperate estuaries, and the causes of this are not yet entirely understood. We do know that seagrass performs an important stabilising role, with rhizomes reducing the erodibility of the seafloor, so any loss of seagrass can lead to an increase in sediment transport and erosion of the adjacent shoreline. This is already occurring along the Adelaide metropolitan shore.

| Coral reefs

Coral reefs cover more than 250 000 square kilometres of the ocean, largely within the tropics. Australia has around 17 per cent of the world's reefs, including the Great Barrier Reef – the largest reef complex in the world – which extends for 2300 kilometres and dominates the northeast of Australia between 9 degrees 30 minutes South and 24 degrees 30 minutes South (see Figure 3.4).

Coral reefs are marine ecosystems, growing on a limestone structure made up primarily of dead coral. Their structure and ecology are linked to the physical nature of the coast and seabed, and the reefs in turn modify waves and constrict tidal energy. They also contribute carbonate sediments to the surrounding seabed and leeward reef flats and islands, called 'cays'. In this section we introduce the distribution of reefs and the composition of reef ecosystems – a more detailed discussion of the nature of reef coasts is given in Chapter 8.

Distribution of coral reefs

Coral reefs occur along much of the tropical shorelines of northern Australia, where sea surface temperatures exceed 18 degrees Celsius in the coolest month (see Figure 2.17). By far the largest expanse of reef occurs along the Great Barrier Reef, and the majority is protected in the largest protected marine area in the world. The Great Barrier Reef Marine Park covers about 345 000 square kilometres of the broad continental shelf, and in places is up to 250 kilometres wide. Nearly 3000 individual reefs have been identified within the park, with a combined surface area of more than 20 000 square kilometres. Coral reefs in Torres Strait, to the north of the Great Barrier Reef, cover more than 37 000 square kilometres, with a further 750 or more reefs around northern Australia covering approximately 6000 square kilometres.

Surprisingly, coral reefs extend into more temperate latitudes; there are well-developed reefs at Middleton and Elizabeth reefs in the Tasman Sea, and the southernmost reefs in the Pacific occur at Lord Howe Island (31 degrees 30 minutes South). There is cover of branching coral (*Acropora*) and turbid water coral (*Turbinaria*) around the five rocky islands forming the Solitary Islands, just offshore from Coffs Harbour (30 degrees South) on the New South Wales coast, and a diverse coral community on the sandstone Flinders Reef near Moreton Bay.

By contrast, the northern and western coasts of Australia have smaller and more isolated reefs. Fringing reefs occur in the lower Gulf of Carpentaria, around parts of Arnhemland, and along much of the Kimberley coast. Fringing reefs extend south along the west coast to Anherst Point at 24 degrees South, where Ningaloo Reef represents perhaps the longest and the most pristine of fringing reefs in the world. The southernmost reefs occur offshore on the Houtman Abrolhos Islands (29 degrees South), which contain some very distinctive coral reefs as well as macroalgae (seaweed) communities. Isolated corals occur at Rottnest Island (32 degrees South) and non-reef forming corals also occur in more temperate waters around southern Australia.

The reefs off Western Australia, although extremely diverse when compared to those in other parts of the world, contain fewer species. The Kimberley reefs are relatively poor in species, but reefs on the continental shelf, such as Ashmore and Rowley Shoals, are diverse (see Table 3.5). Ningaloo Reef and the Abrolhos Islands contain more than 200 species of coral, but diversity drops dramatically south of these reefs. At their more poleward extent, reefs contain fewer species of corals, and there is a greater dominance of coralline algae and other organisms.

Coral reefs also flourish in open tropical waters, and there are reefs on the Cocos (Keeling) Islands and Christmas Island in the Indian Ocean, as well as on Norfolk Island. There are also important reefs in the Coral Sea, and isolated reefs on the extensive continental shelf of northwestern Australia. Mid-ocean reefs such as those on the Cocos (Keeling) Islands contrast with those reefs that occur along the margin of the continent; they receive minimal, if any, runoff from whichever small islands they support, and there is similarly no terrestrial sediment or nutrient input to the reef. A similar situation occurs in the case of the narrow Ningaloo fringing reef, which is backed by Australia's most arid coastal terrain.

Coral reef species

The corals of the world can be divided into two provinces: the Atlantic province, centred on the Caribbean, and the Indo-Pacific province, centred on southeast Asia and northern Australia. Whereas the Atlantic reef province has a relatively limited diversity of scleractinian (reef-building) corals, the Indo-Pacific is extremely diverse. For example, there are three species of *Acropora*, the principal genus of branching coral in the Caribbean, but there are more than 70 species of *Acropora* found on the Great Barrier Reef alone (see Table 3.5).

More than 400 coral taxa are known from the diverse reefs of New Guinea, Torres Strait and the northern Great Barrier Reef. Table 3.5 summarises the distribution of the major families, and within those the most notable genera, of corals, at the major reefs around Australia. It is convenient to consider the distribution of corals, like seagrasses, in terms of those that occur on the west coast and those that

Table 3.5 *Dominant coral taxa: the number of species in the principal families and the most notable genera within them, from key sites around Australia*
Source: Based on Veron, 1995.

Family	Genus	S Western Australia	Rottnest	Abrolhos	Ningaloo	Kimberley	Rowley	Ashmore	Torres Strait	North GBR	Central GBR	South GBR	Eliz Mid	Lord Howe	Solitary Is
Acroporidae	*Acropora*		1	45	37	12	44	48	58	64	72	48	24	7	11
	Montipora	1	1	29	27	7	18	21	25	32	31	24	9	8	6
	other			4	3	1	2	4	7	7	9	6	4	1	
Agariciidae	*Pavona*			5	6	1	8	9	8	8	8	8	5	4	3
	other			7	8		8	10	11	11	11	8	3	2	1
Astrocoeniidae				1	1		2	1	2	2	2	2	2	1	
Caryophylliidae				2	4	4	5	5	7	7	7	6	1		
Dendrophylliidae	*Turbinaria*	3	3	8	9	6	3	5	3	6	9	7	3	4	1
	other			2	2	1			1	2	2	1			1
Faviidae	*Favia*		1	9	8	2	10	12	10	13	13	8	5	3	1
	Goniastrea		2	8	7	4	3	6	5	7	7	7	4	3	2
	Platygyra			2	6	4	4	6	3	4	5	4	3	1	2
	other	2	4	19	24	18	19	30	33	39	38	26	17	9	7
Fungiidae	*Fungia*				5	3	7	10	13	14	14	9	1		
	other			4	9	7	9	14	11	16	18	9	1		1
Merculinidae				3	5	2	3	4	7	7	7	7	2	2	2
Mussidae			2	9	6		7	10	13	15	17	16	8	5	4
Oculinidae				1	2	2	2	2	3	3	3	3			
Pectiniidae				5	8	2	6	11	10	10	9	7	3	2	1
Pocilloporidae	*Pocillopora*		1	4	5	1	3	5	3	1	5	3	1	1	1
	other			2	2	3	2	3	2	4	5	4	2	3	2
Poritidae	*Porites*			9	9	3	8	10	8	2	10	2			
	Goniopora			6	4	6	5	11	12	13	16	8	5	1	2
	other			6	4	2	4	4	3	5	8	3	2	2	
Siderastreidae		2		9	9	6	6	8	9	11	10	6	7	3	3
Trachyphylliidae						1				1	1				
Total		8	15	199	210	98	188	249	267	304	337	232	112	62	51

occur on the east coast of Australia. As can be seen from Table 3.5, the more extensive reefs on the eastern coast, from New Guinea to the Capricorn and Bunker groups at the southern end of the Great Barrier Reef, are rich in corals, with more than 250 species at all except the southernmost sites.

Biodiversity in coral reef ecosystems

Whereas corals provide the framework around which reefs are constructed, coral reef ecosystems are incredibly rich in biodiversity and support a wide range of other organisms, only a few of which produce carbonate skeletons that are preserved. The Great Barrier Reef is justifiably renowned as a centre of marine biodiversity; in addition to the more than 350 species of hard (scleractinian) coral, it contains more than 72 genera of soft coral, more than 300 other species of cnidarian, more than 1500 species of fish, 5000 species of mollusc, 140 species of echinoderm, 1500 species of polychaete worm, 1500 species of sponge, 250 species of ascidian, 500 species of macroalgae, 27 species of marine reptile, 24 species of breeding seabird and 27 species of marine mammal. Ningaloo Reef supports at least 50 species of soft coral and over 520 species of fish, but the richness of the ecosystems supported by other reefs is generally less well known.

Sandy beach ecology

Half of the Australian coast consists of sandy beaches, and as any beach fisher knows, they play host to a range of animals both marine (fish etc.) and terrestrial (birds etc.). What a lot of people do not realise is that beaches and surf also host an entire world of microscopic plankton that lives in the water, and microscopic animals and algae that inhabit the space between the sand grains, as well as numerous species of crustaceans.

While beaches at first appear to be barren and hostile to life, they in fact offer a considerable range and diversity of biotic habitats. Horizontally, these include the subaerial (dunes and upper dry beach), the intertidal (swash zone–shoreline) and the subtidal (surf zone and nearshore), whereas vertically there are pelagic (water column), benthic (surface) and subsurface (interstitial) environments. All are relatively shallow and receive good penetration of sunlight. Most are also exposed to constant water movement, which has the potential to circulate nutrients and exchange water and nutrients with the coastal ocean, inlets and streams. While the above favour a range of habitats and organisms, there are also disadvantages related to the soft, unstable nature of the sand surface, including the potential for burial and erosion, as well as the extreme turbulence generated in parts of the surf and swash zones. As a consequence, flora is restricted to microscopic forms and fauna to highly mobile and specially adapted species. Thus, the physical environment plays a major role in the type and number of species on sandy beaches.

Interstitial organisms live both on and between the sand grains. The empty spaces between the grains can total 40 per cent of beach volume, while the surface area of sand grains increases exponentially with decreasing grain size. The beach and surf zones are occupied by microscopic organisms (<3 millimetres in size) that live both in the pore spaces between the sand grains and on the surface of the grains.

These surfaces host microscopic bacteria, fungi, diatoms and algae. Also occupying the pore spaces are meiofauna (50 micrometres to 1 millimetre in size), including extremely small worms and shrimp; protozoa and small ciliates and foraminifera; and small metazoans, particularly nematodes in finer sediment and copepods in coarser sediments. These organisms may reach an abundance of one million individuals per square metre. They are commonly worm-shaped to assist with movement through the sand grains.

The major form of plant life in the beach system, and the basis of much of the internal food chain, is phytoplankton; that is, microscopic plant plankton that lives in the water column and depends on sunlight for energy. These are commonly called 'surf diatoms' and favour wide dissipative surf zones on gently sloping beaches.

Beach macrofauna

Beach macrofauna consist of those organisms that are too large to move between the sand grains and range from 1 millimetre to 100 millimetres in size. They require a high degree of mobility in this energetic environment and tend to be dominated by mobile filter feeders and scavengers including molluscs, crustaceans and polychaetes, with deposit feeders, particularly crabs, in more sheltered locations, such as tidal flats.

Molluscs are shelled organisms, predominantly bivalves and gastropods. The most common beach bivalve is the species *Donax deltoides* – commonly known in Australia as 'cockles' and 'pipis' – which has a broad global range. Most bivalves are filter feeders, and it is common to find a strong association between high numbers of diatoms and large populations of filter-feeding bivalves. Some gastropods such as *Bullia* are scavengers feeding on carrion. Mussels in general prefer temperate climates, and some colonise moderate to high-energy beach systems, usually buried in the sand with a tube to the surface to filter food from sea water.

Crustaceans have external skeletons and in beach systems include crabs, amphipods, isopods and mysids. These occur both as small meiofauna in the interstitial systems and as larger benthic (bottom-dwelling) and pelagic (open-sea) macrofauna. Polychaetes (worms) tend to occur in the more sheltered locations and most are microscopic in size. However, the giant beach worm (family Onuphidae) is a highly complex group of which there now seems to be a number of species of various genera, lives in energetic swash locations in southern Australia and can grown to more than 2 metres in length.

Fish are the most obvious inhabitants of the surf zone. They support the massive recreational beach fishing industry. While the beach and surf zones are open to many deepwater species, they are dominated by juvenile fishes, which indicates that these zones act as important nursery areas, a consequence of the relative abundance of food and shelter from predators. Fish habitats include both pelagic and demersal (close to sea floor), and benthic feeders include plantivores (feed on plants), piscivores (feed on fish) and omnivores (feed on a range of plants and animals).

Birds, both land and sea-based, are a highly visible part of the beach ecosystem. Birds probably rival the aquatic macrofauna in biomass, with 110 species of seabirds found in Australia and its territories. Seabirds have a significant impact on the ecosystem due to predation and scavenging. Land-based birds feeding intertidally include gulls, sandpipers, sanderlings, snipes, sheathbills, oyster-catchers, plovers and types of cormorants. Most seabirds, including terns, feed predominantly in the surf zone.

The coastal food chain

The beach and surf zones are net producers of nutrients, which are cycled up through a complex food chain. Birds and larger fish make up the top consumers that can then travel, exporting nutrients out of the system into the coastal ocean and beyond.

Australia has a large and rich marine fauna, partly due to its broad latitudinal range. It is estimated that there are 4000–5000 species of fish, with a quarter of these being endemic and most located in southern Australia. It also has at least 166 species of sharks and 177 rays. Two species of seal and one sea lion species breed around the southern coast, with the fur seal inhabiting the beach and dunes at Seal Bay on Kangaroo Island. One herbivorous marine mammal, the dugong (see box 'Seagrass and dugongs', on page 66), feeds on northern seagrass meadows between Shark Bay in the west and Moreton Bay in the east.

Many of Australia's beaches contain the broken-up detritus of shells and other organisms in their sediment (see Chapter 5). In some cases up to 100 per cent of the beach is composed of marine material, and in these areas the beach and its ecosystem are inextricably linked.

| Rocky shore ecology

Much of the other half of the Australian coast is dominated by rocky shores, which provide a very different habitat compared to beaches. Rocky shores are usually even more exposed than beaches with waves breaking heavily on the more exposed locations, while they can also provide sheltered habitats like tidal pools. The major difference, however, is the fact that rocky shores provide a hard, stable surface, and no matter how exposed to breaking waves, there are organisms that thrive in this environment. They also provide a range of habitats from the upper level of wave spray and swash, across the intertidal to the subtidal. These zones are controlled by the regular tidal movements, together with variable wave and wind conditions, which can range from calm to storms, as well as varying between exposed locations and sheltered bays. Their rock surfaces can range from vertical to horizontal in gradient. The result of all this is a complex, although reasonably predictable arrangement of supratidal, intertidal and subtidal organisms. Like coastal dunes, a number of species have adapted to this environment and, like the dunes, are arranged in rough zones between the upper limit of wave swash and spray to the subtidal. In general, rocky coastal ecology varies at three scales: first, shore normal from subtidal to the upper subaerial limit; second, between sheltered and exposed locations, and third, between different biogeographic regions around Australia.

In temperate Australia, the typical gradation from the upper to lower rocky shore begins with the small blue periwinkles (*Littorina* spp.) and limpets (for example, *Notoacmaea petterdi*), which occur above sea level in a usually dry zone that extends to the upper reaches of the swash and spray. Some can be found up to 12 metres above sea level. These grade into the barnacle zone, which is usually located between the high water mark and mid-high tide. Species include the small *Chamaesipho columna*, the slightly larger *Chthamalus antennatus* that usually marks the upper limit of barnacles, and in exposed locations the surf-barnacles (*Catophragmus polymerus* and *Tetaclita rosea*). At the mid-tide level, species are in part dependent on the level of wave action, which varies from sheltered locations favouring oysters (*Saccostrea* sp.) and the hairy mussel (*Brachidontes erosus*), to the open coast where grazing snails (*Austrocochlea constricta* and *Bembicium nanum*) and limpets dominate.

Below high tide begins a zone of hard, white tube-worms, dominated in New South Wales by *Galeolaria caespitosa*. It is restricted to locations exposed to heavy surf and will reach higher elevation in locations reached by swash. The lowest of the intertidal zones is occupied by cunjevoi (*Pyura stolonifera*), which extends from the lower intertidal down into the subtidal. This zone also contains the grazing chitons (*Plaxiphora albida* and *Onithochiton quercinus*). The subtidal, which is never exposed, is dominated by dense covers of seaweeds (macroalgae), including *Corallina officinalis*, brown kelps (*Ecklonia radiata*), and in southern Australia bull-kelps (*Durvillaea potatorum*) and the long brown algae *Sargassum* spp., together with an abundance of sessile and mobile invertebrates (that is, sponges, bryozoans, ascidians, hydroids, enchinoderms, molluscs and crustaceans).

Some other notable animals found on rocky shores and in tidal pools are the starfish, including the larger *Patiriella calcar* and smaller *Patiriella exigua*, and the spiky sea-urchins (*Centrostephanus rodgersii*), which do live in deeper tidal pools, but are more abundant in the subtidal.

The subtidal temperate rocky reefs are characterised by large macroalgae and an abundance of sessile and mobile invertebrates. Little is known, however, of the ecology of tropical rocky shores and subtidal rocky reefs.

Conclusion

The coastal landscapes shaped by the processes described in Chapter 2 and the adjoining water bodies provide habitat for a wide range of terrestrial and marine ecosystems, whose nature and extent are also influenced by air and ocean temperatures and on land by precipitation. These include coastal dune vegetation, salt marshes, mangroves, seagrasses and coral reef systems.

Australia has a physically long coast with a range of tropical through temperate climates, each with its own characteristic set of species and ecosystems. The tropics play host to the world's most extensive coral reef systems, which are described in more detail in Chapter 8, as well as the third-largest area of mangroves in the world. Tropical seagrasses fringe the entire north coast, while temperate seagrasses reach their greatest extent on the planet around the southern coast.

Our rocky shores and beaches support a surprisingly wide range of often not-too-visible flora and fauna, whereas the backing coastal dunes are well vegetated by world standards. In part, this is a reflection of the adaptation of a range of tropical and temperate grasses, shrubs and trees to their infertile environment. The rocky shores, beaches and coastal dunes host broader ecosystems, each containing a wide range of flora and fauna that has adapted to that particular environment and in total ensure a distinctively Australian coast.

ESTUARIES AND DELTAS

Introduction

ESTUARIES AND DELTAS occur where creeks and rivers meet the sea. There are more than 1000 rivers and estuaries around the coast of Australia, together with over 1300 smaller creeks that drain to the coast. The availability of a navigable river with sufficiently deep water for shipping led to the initial establishment and growth of Australia's early settlements (Sydney Harbour, Hobart on the Derwent River, Brisbane on the Brisbane River and Perth and Fremantle on the Swan River, discussed further in Chapter 9). Some, such as Sydney Harbour and adjacent Botany Bay, have become the focus of major cities, and are the location of busy ports. However, today more than 50 per cent of Australian coastal waterways remain remote from settlements and in near-pristine condition.

There are large deltas actively building seaward at the mouths of many of the world's largest rivers, such as the Nile (North Africa), Mississippi (North America), Mekong (Indo-Chinese Peninsula) and Ganges–Brahmaputra (South Asia) rivers. Not surprisingly, all the world's major deltas are located in the humid tropics to subtropics, where the catchments receive copious rainfall and deliver large volumes of water and sediment to the coast.

Australian deltas

Although northern Australia is located in the tropical and subtropical latitudes, its deltas are not large by world standards. There are several reasons for this. First, Australia is the driest of the inhabited continents (Antarctica actually receives less

ESTUARIES AND DELTAS

Estuaries and deltas occur at the seaward end of large creeks and river systems, where they are influenced by interaction with marine processes. The term **estuary** is derived from the Latin *aestus* ('tide'), indicating that tidal processes contribute to estuarine morphology and drive estuarine processes. Estuaries include a diverse range of coastal waterways, from large drowned valleys, such as Sydney Harbour, and embayments, to smaller intermittently closed and open lakes and lagoons, and the mouths of all rivers.

The term **delta** comes from the Greek letter *delta*, which is triangular and was used to describe the (deltaic) mouth of the river Nile, which splits into several distributaries in the form of a triangle before it empties into the Mediterranean Sea.

At the simplest level, estuaries tend to be indented waterbodies, where marine influence penetrates inland, whereas deltas protrude seawards as creeks and rivers enter the sea and deposit sediment. However, estuaries may contain deltas, and the mouths of delta distributaries behave as estuaries.

precipitation). Second, a large proportion of Australia's interior has either no drainage, or an internal drainage system (see Figure 4.1). For example, about one-sixth of the inland drains into Lake Eyre, which lies at the centre of the continent at an elevation of about 15 metres below sea level. Third, many Australian rivers are periodically dry or highly irregular in their flows, and flows are highly episodic. In addition, rivers across northern Australia are dominated by the wet summer monsoon, and discharge is extremely seasonal, with little to no flow during the long dry season. Australia is also a low-lying continent composed of old and weathered rock. In a global context it has a low rate of soil erosion and sediment loss. All these facts combine to produce a continent with only a few seasonally large rivers (such as the Ord, McArthur, Burdekin and Fitzroy rivers), numerous smaller, intermittent to dry river courses and relatively few deltas.

Even the country's largest river system, the Murray–Darling river basin, brings modest discharge and sediment loads to the coast, and remains a highly modified estuary rather than a delta, at its mouth. As a consequence, deltas are less prominent in Australia than on most continents. In this chapter we use the term 'delta' more broadly in an Australian context to describe estuaries that have reached that stage of infill where they are supplying sand to adjacent coasts.

Understanding estuaries

Estuaries represent a trap for sediment. They are relatively sheltered coastal embayments (drowned valleys, bays and lagoons that act as a sink for terrestrial sediment delivered by creeks and rivers). Estuaries also accumulate marine sands delivered through their entrances by wave and tidal processes, and are sites for accumulation of locally produced carbonate sediments, such as shell. Terrestrial sediments range from gravel, as in the Pioneer River, Queensland, to sand, which dominates the majority of catchments, to finer silt and clay, as in the southern Gulf of Carpentaria.

The sediment derived from the coastal catchments is primarily angular quartz sand, gravels and other lesser minerals and mud. The sand derived from seaward, delivered by waves and tidal flow, contains more rounded quartz grains with shell hash and other carbonate material. On wave-dominated coasts, waves and tidal currents build massive flood-tide deltas into coastal inlets. Observations of the stratigraphy that flanks large macrotidal estuaries in northern Australia – that is, the sequence of sediments that underlie them – suggest that tides have gradually delivered sediments to infill these systems.

More immediate evidence for the effectiveness of tides in bringing mud into estuaries comes from infill of the lower reaches of the Ord River in Western Australia since completion of the Lake Argyle dam in 1972, which decreased river flow and permitted increased tidal influence and consequent sedimentation.

In order to understand the rich diversity of estuarine landforms, we need to understand the factors that determine the stage of estuarine infill, as well as the relative balance between wave and tide processes. Patterns and rates of infill are slow, generally imperceptible to the observer, but we can predict these processes using studies of the stratigraphy and chronology of infill. We are also able to gain insights by comparing adjacent systems, which are at different stages of the evolutionary sequence. Many of the wave-dominated estuaries and lagoons along the southwest and southeast margins are still largely unfilled and continue to trap river sediment, resulting in little if any supply of river sand to adjacent coasts. Some, such as Sydney Harbour and Jervis Bay, represent enormous basins that are only receiving sediment from small rivers or creeks, and are therefore still deep and flanked with bedrock. Others, such as the Richmond and Shoalhaven rivers in New South Wales and many of the rivers in northeastern Queensland, have largely infilled their estuaries and now supply river sand to adjacent coasts. These are more appropriately considered wave-dominated deltas, as discussed below.

Types of estuaries and deltas

There have been several attempts to classify Australian estuaries based on factors such as climate,

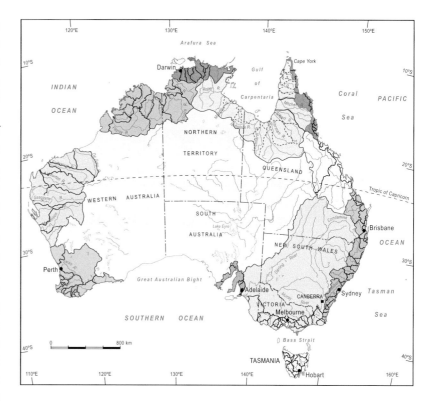

hydrology, water quality and ecology. However, a geomorphological classification provides a useful framework for describing the landforms, as well as the structure and functioning of estuarine ecosystems.

A river that enters a lake or lagoon will flow into that body of still water as a jet. As the rate of flow slows, the coarser sediment (gravel, if present and coarse sand) is deposited first as a bar at the river's mouth. Then at greater distance into the lake, fine sand and eventually mud (comprising silt and clay) settle out more slowly on the flat floor, forming what is called a 'prodelta'. Gradually, a sequence of layers of sand, called 'foresets', are deposited, dipping into

Figure 4.1 *The distribution of major catchments that drain to the coast across Australia, showing the larger rivers. All of central Australia and parts of the northwest and south coast have no drainage, or are dominated by internal drainage.*

the lake. In this way, the sandy delta builds out (or progrades) over the deeper, more horizontal and muddy prodelta. When a river discharges into the sea, the depositional processes are influenced by two factors that are generally absent or imperceptible in enclosed water bodies: waves and tides.

Coastal waterways can be classified in terms of the relative dominance of river, wave and tidal processes, as shown in Figure 4.2. Only a few of the larger systems can be regarded as river-dominated, because Australian rivers do not carry sufficient flow or sediment loads. Instead, the triangular diagram in Figure 4.2 is divided symmetrically into wave-dominated systems (deltas and estuaries, lagoons and beach-ridge plains), in which waves are the dominant process, and tide-dominated systems (deltas, estuaries, tidal creeks), in which tides are dominant. The wave-dominated systems have shore-parallel, wave-built landforms, whereas the tide-dominated systems have predominantly shore–normal features aligned perpendicular to the shore and shaped by tidal flows.

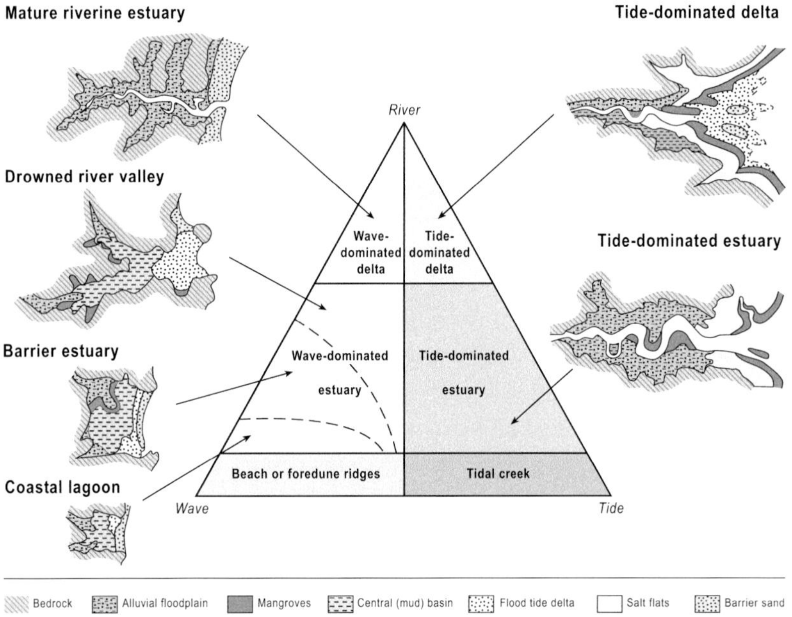

Wave-dominated systems

Wave-dominated estuaries and deltas occur around the wave-dominated southern Australian coast. Their entrances are characterised by wave-built, shore-parallel features, such as beaches, barriers and spits. Figure 4.3 shows waves at the mouth of the largest catchment, that of the Murray–Darling system, where waves rework the river-mouth sand to narrow, and on occasions close, the entrance even of this mighty river. The mouths of rivers draining smaller catchments are shaped primarily by waves if they are exposed to significant wave energy. For example, the river may be diverted several kilometres in the direction of longshore drift, as seen in the case of the Henty River in western Tasmania (see Figure 4.21). As the river supplies more sand, wave action may build a series of shore-parallel beach or foredune ridges (see Chapter 6).

On the wave-dominated coast of southeastern Australia, estuaries can be divided into three main types, and these are shown schematically in Figure 4.2:

1. **Coastal lagoons** are predominantly closed off from the sea by a sandy barrier. They occur where discharge from a small river, or more often a creek, is insufficient to maintain a permanently open inlet.
2. **Barrier estuaries** occur where a sandy barrier separates a large waterbody from the sea,

Figure 4.2
Representation of Australian estuaries and deltas in terms of the relative dominance of river, wave and tidal processes. A wave-dominated delta, in the Australian context, corresponds to a mature riverine estuary. A wave-dominated estuary in Australia includes a drowned river valley, barrier estuary or coastal lagoon.

but limited tidal exchange occurs through a constricted inlet.

3. **Drowned river valleys** are deeper, bedrock-fringed embayments, with a submerged tidal delta at their usually wide, deep mouths.

Tide-dominated systems

The coast of Australia has over 1000 small tidal creek systems, and these are particularly abundant across the broad, saline mudflats of northern Australia and down the eastern seaboard. A typical northern tidal creek, such as shown in Figure 4.4, has a branching pattern that is called 'dendritic', since it resembles the branches of a tree. The tide shapes creeks so that their width tapers rapidly upstream, as a result of a reduction in tidal prism (the volume of water that comes in on the flood tide and leaves on the ebb tide). Creeks adopt a meandering form, frequently with mangroves lining their banks. Saltwater floods out over the creek banks at high tide and inundates the muddy surface of the tidal flats. It subsequently drains into the creek again on the ebb tide, leaving a thin drape of mud across the surface, and often a thin crust of salt.

Larger river mouths that are dominated by tides form tide-dominated estuaries and deltas. These are generally funnel-shaped, with extensive muddy tidal flats and features such as elongated intertidal or subtidal sand banks, which parallel the tidal currents. The many large, mangrove-fringed, tide-dominated estuaries along the northern coast of Australia include the Ord, Keep, Victoria, Finiss and Adelaide rivers, where tidal range is more than 5 metres. On the northwestern megatidal coast, the estuaries have tides of up to 12 metres in amplitude, such as at King Sound (Fitzroy River).

Figure 4.3 *Characteristics of typical wave-dominated systems, showing shore-parallel features at the mouth of the Murray River in South Australia. Although this is the largest of the catchments draining to the coast, there is now very little flow actually reaching the mouth, and the considerable wave power has on occasions closed the mouth.* Photo: A.D. Short

Figure 4.4 *A typical tidal creek in tropical northern Australia, showing the meandering channel and branching (dendritic) drainage pattern. There is a prominent mangrove fringe along the coast, which continues up the major creeks.* Photo: A.D. Short

Tide-dominated deltas are similar to tide-dominated estuaries; the term is used in Australia to refer to systems where the estuarine basin (the accommodation space) has been largely infilled and the coast is building seaward, as represented by sequences of beach ridges or cheniers (see Chapter 6).

River-dominated systems

Whereas those large tropical and subtropical deltas of Asia and elsewhere in the world are clearly shaped by rivers that bring vast quantities of sediment to the coast, river domination is not such a common feature of Australian waterways. Here, waves tend to dominate the southern coast, and tides the northern coast. Only inside estuaries – away from waves and where tidal range is limited – do creeks and rivers build the classic river-dominated delta. In their simplest form these fluvial deltas consist of sinuous river levees that protrude into the bay or lagoon, in some cases bifurcating into two or more distributaries. Examples include Dora Creek, which has built levees into Lake Macquarie, Tuggerah Creek, which extends into Tuggerah Lake, and Macquarie Rivulet, which has built a branching delta into Lake Illawarra (see Figure 4.6).

Rivers also exert an increasing influence on coastal waterways as estuarine embayments infill and develop into more delta-like forms. This later stage occurs when the river channel crosses the broadly infilled former estuary, and as successive floods build levees on either side of the channel, separating the river from the broad floodplain on all but the largest of overbank floods. This stage results in a 'mature riverine estuary' in wave-dominated systems, or as it is frequently called, a 'wave-dominated delta'. In this situation, the river supplies sediment directly to the sea, as opposed to earlier stages, when most sediment is sequestered within the estuary.

Distribution of estuaries and deltas

The framework in Figure 4.2 suggests that each system can be classified according to its dominant process: waves, tide or river. At a broad level, it is possible to define where each waterbody would be positioned. The relative contribution of wave and tidal power could be expressed using a measure such as the relative tidal range (RTR) (see Chapter 5), and river flow, which is measured as mean annual discharge, could give a relative measure of the importance of the river in shaping its mouth. In practice, however, these parameters are generally poorly understood. They vary both over time (particularly for the systems that are influenced by seasonal or monsoonal variations in climate) and along the length of many coastal waterways. Nevertheless, there have been a few classifications of estuaries around Australia, and preliminary attempts to describe their geomorphological characteristics and functioning.

An initial Australian Estuary Database (AED) formed the basis for an assessment under the National Land and Water Audit. This was later expanded to form an internet-based database of coastal waterways available through a Geoscience Australia portal, called OzEstuaries. Recently, OzEstuaries has been incorporated into a larger coastal database, called OzCoasts. The Australian Beach Safety and Management Program (ABSMP) independently compiled a database on the entire open Australian coast, and it also has produced further data on coastal waterways.

How the systems differ

Figure 4.5 shows the geographical distribution of estuarine and deltaic systems around Australia, based on mapping undertaken as part of OzEstuaries. Table 4.1 summarises the state-by-state distribution of the principal types of system, based on some of the early mapping. Criteria for discriminating one type of system from another are not clear cut, particularly since many coastal waterways contain parts that appear more estuarine and others that appear more deltaic, and river influence varies both geographically within a system and over time. Nevertheless, some broad trends are apparent. There is a predominance of tide-dominated systems in northern Australia and wave-dominated systems in southern Australia. The areas of large tidal range in the Northern Territory and northern Western Australia contain numerous tide-dominated estuaries, together with tide-dominated deltas in the Gulf of Carpentaria and in sheltered bays along the east Queensland coast. Wave-dominated estuaries are especially prominent in New South Wales, Victoria and southwest Western Australia, whereas wave-dominated deltas, reflecting a more mature stage of infill, are more characteristic of northern New South Wales and Cape York Peninsula.

The classification shown in Figure 4.5 is a broad generalisation. It shows the relative balance of wave and tidal processes, which is what is significant, rather than their absolute value, since the relative dominance of processes can change over time. For instance, in tropical Australia there is a

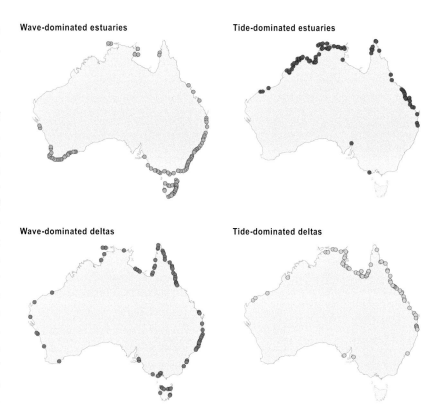

pronounced monsoonal influence that causes river flow to dominate in the wet season and tidal flow to dominate in the dry season. Also, many rivers are influenced by extreme events. In Australia, which has relatively limited precipitation and low sediment loads, there are few systems that are perennially dominated by river influence throughout the year, or that would be described as deltas in the usual sense of the term. Nevertheless, it seems appropriate to place those wave-dominated or tide-dominated systems that have infilled the prior embayment, are supplying sediment to the sea and prograding at the coast, into the delta field of Figure 4.2.

Figure 4.5 *The distribution of wave-dominated estuaries, tide-dominated estuaries, wave-dominated deltas and tide-dominated deltas around Australia.* Source: based on Heap, Bryce, Ryan, Radke et al., 2001 and data in the OzCoasts database, <www.ozcoasts.org.au>

Waterway type	Qld	NSW	Vic	Tas	SA	WA	NT	Total
Wave-dominated delta	44	18	5	10	2	7	7	93
Wave-dominated estuary	8	57	21	32	2	31	6	157
Tide-dominated delta	50	1	0	1	1	4	16	73
Tide-dominated estuary	38	0	1	3	1	24	28	95
Beach foredune ridges	19	9	10	5	0	5	12	60
Tidal creek/tidal flat	140	3	2	3	10	73	54	285
Total	299	88	39	54	16	144	123	763

Table 4.1 *Australian estuaries and coastal waterways, indicating number in the major geomorphological types by state*

The OzEstuaries database considered more than 700 waterways around the Australian coast, including 37 per cent that are classified as tidal creeks and 8 per cent classified as beach foredune-ridge plains (also called 'strandplains') (see Table 4.1). Of the estuarine types, wave-dominated estuaries are most common, accounting for 21 per cent, whereas tide-dominated estuaries account for 12 per cent; wave-dominated deltas account for around 12 per cent and tide-dominated deltas less than 10 per cent, with the remainder not easily accommodated within the classification. The ABSMP database indicates that there are over 1200 creeks, 700 small rivers and 200 larger rivers reaching the coast. Of these, 50 per cent are located across the humid, tropical coast of northern Australia, 30 per cent in the humid temperate southeast (New South Wales, Tasmania and Victoria) and only 30 per cent along the long, but more arid coast of South Australia and Western Australia as far as Broome. These systems contain 1320 estuaries (60 per cent). Of these, 66 per cent are wave-dominated barrier estuaries, with 22 per cent saline coastal lagoons and 12 per cent drowned river valleys.

The coast of southeastern Australia provides a classic example of a wave-dominated estuarine coast, together with wave-dominated beaches (described in Chapter 5). The macrotidal coasts of northern and northwestern Australia, on the other hand, contain tide-dominated, funnel-shaped estuaries, with muddy banks and linear shore-normal sand shoals. More than 60 per cent of Australia's deltas and 50 per cent of the prograding plains and tidal flats occur on the northeastern margin, reflecting the influence of the monsoonal climate. In terms of estuary evolution, ABSMP indicates that 30 per cent are completely infilled and essentially a delta, 52 per cent are substantially infilled, 14 per cent about half infilled, with the remaining 4 per cent only partly infilled.

Estuarine habitats

Estuaries and deltas occupy the transition from freshwater to marine. In arid regions they also contain hypersaline habitats. The ecosystems comprise species that are able to survive across gradients in salinity, temperature and turbidity (the varying amount of fine sediment in suspension in the water). Estuaries offer a shallow, stable, protected coastal environment that receives nutrients from the land and sea, is flushed twice daily by tides and generally has direct access to the sea. Although they don't contain many species, estuarine habitats are often particularly productive.

In estuaries and deltas, the rate of primary production, the fixation of the sun's energy through photosynthesis, is high. This high rate of production is achieved by plants, including microscopic plankton, seaweeds (algae), seagrass and other

macrophytes such as mangrove, saltmarsh and freshwater wetland vegetation (as discussed in Chapter 3). Estuaries can maintain this productivity because they receive large inputs of nutrients from both rivers and the sea. These plants provide direct nutrition for herbivores, including wildfowl such as black swans, which feed directly on plant material, and dugong and turtles, which graze on seagrass.

However, still more important is the detrital food web. In the case of seagrass, there is often a rich community of algae and micro-organisms on blades of seagrass (termed 'epibiotic'), and these together with other plant debris, such as mangrove leaves, form a substrate that is colonised and re-colonised by micro-organisms. Thus, although the macrophyte detritus is usually slow to break down and low in

Figure 4.6 *The fluvial delta of Macquarie Rivulet has built into Lake Illawarra and is prograding over the mud of the central basin.*
Photo: M. Robinson

nutrients itself, it is consumed repeatedly by a range of organisms, such as worms, crustaceans, and fish, which in turn are the food for larger fish and birds. Consequently, estuaries support major estuarine food webs that link directly into marine ecosystems in shallow nearshore waters, and they also contribute to terrestrial food chains through birds and insects.

The morphology of an estuary depends on the variety of processes that operate along its length and over time as sediment builds up. Understanding how these systems function is important if we are to effectively manage our significant estuarine resources.

Wave-dominated estuaries

Three distinct sedimentation zones can be identified in wave-dominated estuaries: the barrier and inlet, central basin and fluvial delta. To seaward, land-forms are built with barrier sands, predominantly with marine sand deposited by wave action as beaches and, in most cases, topped by dunes forming a coastal barrier as described in greater detail in Chapter 6. The influence of the river increases to landwards, across a low-energy waterbody, the central basin, which is characterised by deposition of mud supplied by the river. The river deposits its coarse load (gravel and sand) as its flow decelerates and builds a fluvial delta at its mouth (also called a 'bayhead delta'), which progrades over the mud deposited in the central basin. The accumulation of these sand and gravel deposits over the mud infills the central basin.

Delta distributaries can bifurcate to form a digitate, or birds-foot, delta, often leaving sheltered, cut-off embayments (re-entrants) in which mud is deposited (see Figure 4.6). The delta may split into several distinct distributaries, forming a river-dominated delta. This is a delta building into an estuary, illustrating the problems in distinguishing between deltas and estuaries. Over time, the mud fills the central basin and the sandy delta builds out over the shallow mud-basin surface to form a floodplain that eventually could completely cover the basin and fill the estuary. Thirty per cent of Australia's estuaries have reached this stage.

Habitat characteristics

The three sedimentation zones also provide three distinct habitats. The sand barrier consists of the beach, usually backed by a vegetated dune. The adjacent inlet contains intertidal and subtidal sand bodies, which are reworked by tidal energy and waves to form submerged flood tide deltas. Tidal deltas are intertidal and subtidal lobes of sand in tidal inlets, constructed by ebb or, more commonly, flood tides. A flood tide delta is formed by wave action and tidal flows through an inlet, which deposit a body of clean marine sand that is distinctly different from the fluvial delta.

The central basin is the open estuarine waterbody, commonly a lagoon, in which mud accumulates. River flow rapidly decelerates here, but there is restricted tidal variation and the basin may be fringed by low-gradient, fine-grained intertidal flats colonised by seagrasses, saltmarsh or mangroves.

The fluvial delta builds into the central basin. As the river channel extends into the estuary, it is generally flanked by sandy or silty levees. Behind the levees, there are often substantial areas of floodplain, particularly in those systems that have reached a more mature stage of infill. The floodplain, as its name suggests, may be flooded intermittently, and wetlands can persist where the natural ecosystems have not been cleared for agriculture. The saltmarsh that is so typical of those areas where tidal influence is still felt, merges into broad-leafed paperbark (*Melaleuca* spp.) and Sheoak (*Allocasuarina glauca*), and wetlands with sedges, reeds and rushes.

Components of this tripartite division can be seen in each of the types of wave-dominated systems shown in Figure 4.7. Waves are effective in building a sandy barrier where the nearshore is shallow, but wave energy is attenuated into the entrance of drowned river valleys, where wave and tidal processes move sand into large submerged tidal deltas; in this case, the seafloor is too deep for these sands to form a barrier above sea level. The edges of these largely submerged tidal deltas reach the surface as estuarine beaches and spits, as in Sydney Harbour at The Spit, in Port Hacking at Deeban Spit and Port Stephens along Winda Woppa Spit.

River and tidal influences

Typical gradients in wave, tide and river power are shown in Figure 4.7. Wave power is at a maximum at the seaward end. It is reduced by the narrow inlet in barrier estuaries and lagoons, but can penetrate into the wider, deeper mouths of the drowned river valley. On larger estuarine waterbodies, wind-waves are generated across the central basin. These small waves can shape the shoreline, and it is not unusual for the shoreline of barrier estuaries and coastal lagoons to consist of low estuarine beaches and recurved sections where sand is moved by longshore drift,

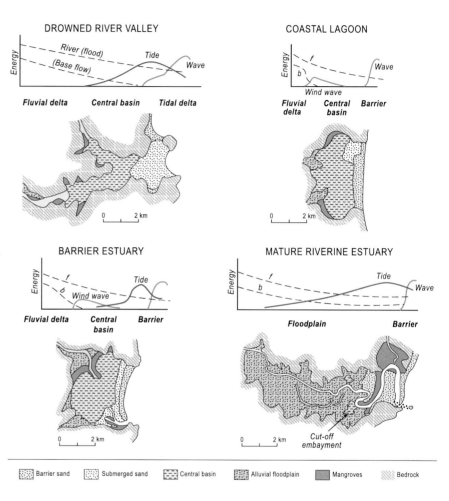

Figure 4.7 *The main types of wave-dominated estuaries, and a schematic representation of the distribution of energy along each estuary type. River input is generally highly episodic and both flood (f) and baseflow (b) conditions are indicated. Note that coastal lagoons may breach, in which case they behave like barrier estuaries.* Source: based on Roy et al., 2001

between rocky outcrops. River flows dominate the river section and fluvial delta, then decrease rapidly beyond the fluvial delta, except during exceptional flood events. Only in the case of the mature riverine estuary does the river maintain a strong influence throughout.

Tidal range, and consequently tidal influence, differs amongst the different types of wave-dominated systems. In a drowned river valley, the tidal delta is generally completely submerged and the tidal range is similar to that in the open ocean. Tides are attenuated in a constricted barrier estuary; tidal processes are strongest at the entrance, and their influence can extend up the estuarine embayment depending upon its morphology. In the case of barrier estuaries, a flood tidal delta is deposited as a lobe of marine sands worked into the estuary. Less frequently, there may also be an ebb tidal delta, built by the outgoing tide. There is usually no tidal variation in coastal lagoons when they are completely closed.

Estuary evolution

Estuaries around the Australian coast have been infilling with sediment from one of three sources: from the land (terrestrial), seaward (marine) or produced within the estuary (in situ). Infill began about 6000 years ago, when sea level reached a level close to its present around the entire coast (see Figure 1.7). Sea level reached 1–2 metres above its present in some places, followed by a period of fall to present levels, but this slight change is not discernible in the pattern of estuarine sedimentation. Over the past 6000 years, estuaries have continued to trap sediment, but about 70 per cent of the wave-dominated estuaries around the Australian coastline remain only partially infilled. Australia has experienced little vertical movement from tectonic or isostatic activity, in contrast to those areas that were covered by, or close to, the polar ice caps (see Chapter 1). As a consequence, most Australian wave-dominated estuaries are still infilling and do not supply sediment, other than mud, directly to the coast. The time required to completely fill the estuary, so it transforms to a delta supplying sand directly to adjacent coastlines, depends on the volume of the estuarine accommodation space (the initial topographic depression in which the sediment infills), and the rate of sediment supply. Only 30 per cent of Australian estuaries have achieved this after 6000 years of infilling, with the remainder requiring a longer period of time.

The morphology of each system depends on the pre-existing topography and the contemporary processes that supply the sediment. Inheritance from previous cycles of sea-level change can also be important. Estuarine sedimentation also filled the topographic depressions that form estuaries during former sea-level highstands, particularly the Last Interglacial (around 125 000 years ago). However, estuarine sediments are relatively easily eroded as a result of incision by rivers at times when sea level is lower. The extent of unconsolidated or weakly lithified Pleistocene sediment and the degree to which it influences the geomorphology of modern estuaries differs between individual systems. For example, an 'inner barrier' of Pleistocene age (generally Last Interglacial, although perhaps incorporating sediments deposited during previous highstands) has been well preserved in northern New South Wales barrier estuaries, but is only intermittently

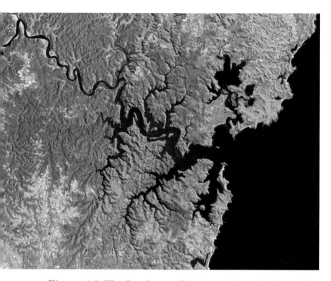

Figure 4.8 *This Landsat satellite image shows the drowned river valley of the Hawkesbury River, one of several dendritic estuaries in the Sydney region. The river incised into the landscape when sea level was lower, and these former river valleys have been 'drowned' by the rise of sea level following the last glaciation.* Image: © Commonwealth of Australia, ACRES, Geoscience Australia

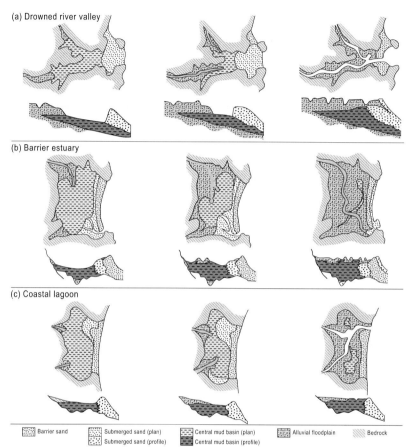

Figure 4.9 *The three stages of infill of each of the wave-dominated estuaries:* **a)** *Drowned river valley,* **b)** *Barrier estuary,* **c)** *Coastal lagoon.* Source: based on Roy, Williams, Jones, Yassini et al., 2001

encountered, usually at depth, in southern New South Wales estuaries. The inner barrier and outer barrier give Myall Lakes their distinct appearance (see Chapter 6), with the modern Lower Myall River occupying the depression between the two barriers.

Drowned river valleys

The pattern of sedimentation and change in estuaries is relatively slow. The pattern can best be determined from coring and dating of sediments deposited within the estuary or in the extensive plains that represent formerly infilled estuarine areas. This procedure has been particularly effectively undertaken for southeastern Australia. Drowned river valleys are essentially bedrock-flanked embayments. Sydney Harbour (Port Jackson) is a classic example; it receives relatively little river input because the Parramatta River has only a small catchment, and as a consequence the harbour is only partially infilled and remains deep and flanked by bedrock. Hawkesbury sandstone forms the prominent cliffs at the mouth of the embayment (North Head and South Head), and around much of its margin. Port Jackson, as well as Port Hacking and Broken Bay at the mouth of the Hawkesbury River, are all drowned river valleys, and their numerous arms can be more clearly understood as part of the dendritic drainage pattern of former river systems drowned by sea-level rise (see Figure 4.8).

The Hacking River and other tributaries have supplied only a little sediment to the large embayment of Port Hacking. The estuary is tidal and saline throughout, except during major floods. The central basin has a muddy floor at depths of up to 20 metres. A large flood-tidal delta of clean marine sands with some shell extends several kilometres up the embayment. Radiocarbon dating has shown that this has built steadily further into the embayment over the past 8000 years (see Figure 4.9a). The Hacking River has built a small fluvial delta where it discharges into the estuary, with slower rates of alluvial infill in tributary embayments where streams flow in, such as at Gunnamatta Bay. The sequence of infill by tidal delta and fluvial delta is shown schematically in Figure 4.9 – marine sand deposits build into the central basin, burying the terrestrial mud.

Barrier estuaries

Barrier estuaries comprise a sand barrier that separates the estuarine water body – the central basin – from the sea. Lake Macquarie and Lake Illawarra are two of the larger examples along the New South Wales coast, while Narrabeen Lagoon offers the best example in the Sydney region. The barrier formed as a result of landward migration of sand during postglacial sea-level rise; the sands had previously been carried from the catchment onto the shelf. The central basin was partially occluded behind the barrier as it was reworked landward. This central basin is infilling with mud (silt and clay) supplied by the incoming streams, with the streams building fluvial deltas where they flow into the lake (see Figure 4.9b). In some cases, such as Macquarie Rivulet building into Lake Illawarra (see Figure 4.6), the river may split into several distributaries, called a 'birds-foot delta' (because channels split, or bifurcate, forming a shape rather like the foot of a bird). The flow decelerates and the coarser material is deposited first, because the velocity is no longer sufficient to keep grains in motion. In other situations, an elongate levee is constructed, as where the Mitchell River builds silt jetties into Lake King in the Gippsland Lakes of Victoria.

Infill of barrier estuaries

Barrier estuaries also receive limited sediment from seaward, as indicated in Figure 4.9b. The inlet is dominated by strong tidal flows, as indicated in Figure 4.7, which transport marine sand into the lagoon. This builds a flood tide delta and also adds sediment to the sandy back barrier. The central basin is a sheltered body of water, although in the larger systems wind-waves generated across the surface of the lake are important in reworking shallow basin sediments, to maintain a minimum depth, and sculpting estuarine beaches and spits. Vertical sedimentation rates in the centre of the lake are gradual – for example, in Lake Illawarra the rate is about 1 millimetre a year on average.

The rate of infill mostly depends on the size of the catchment and the volume of sediment supplied, relative to the pre-existing basin topography. As a consequence, we are able to envisage a sequence of stages of infill, reconstructed from the stratigraphy of the more mature systems, or inferred by comparison with adjacent systems that are at different stages of the continuum. Some barrier estuaries remain as deep basins (for example, Burrill Lake, Lake Conjola; see Figure 4.10), whereas others at more mature stages of infill are shallow and swampy around their margins (for example, Tabourie Lake). Ultimately, the majority of accommodation space in the central basin will infill, and the river channels will then meander through a largely infilled floodplain, leaving a few remnant, cut-off embayments generally fringed with mangroves. Alluvial floodplains form over the previously subtidal or intertidal environments, and a mature riverine estuary is formed.

Coastal lagoons

Coastal lagoons are usually small in area and are fed by creeks rather than rivers. They occur at the mouths of smaller catchments, and many of these only open to the sea intermittently during floods. When they are closed, the salinity of coastal lagoons can increase through evaporation, particularly in arid regions. However, it can also be reduced by

freshwater inflow in more humid areas. When lake levels rise they trigger the opening of the lakes at the lowest point along the sandy barrier, although nowadays many are opened mechanically by local councils, to reduce flooding and improve water quality.

Infill of coastal lagoons

Infill of coastal lagoons follows a similar pattern to that of barrier estuaries (see Figure 4.9c). A mature riverine plain is the likely ultimate stage, although this has rarely been reached by these smaller systems. The Tuross River and Coila Lake, on either side of the peninsula that links to Tuross Heads (see Figure 4.11), illustrate the contrast between a barrier estuary and a coastal lagoon. The Tuross River maintains its flow and hence its mouth has remained

Figure 4.10 An aerial view of a barrier estuary, Lake Conjola, in southern New South Wales. The inlet may close through wave action after periods of prolonged low river flow, but is generally open. Photo: A.D. Short

Figure 4.12 *Wamberal Lagoon, pictured here with its central basin area of 0.23 square kilometres, is one of a series of coastal lagoons in central-southern New South Wales which is more frequently closed than open to the sea.* Photo: A.D. Short

Figure 4.11 *A Landsat satellite image showing contrasting coastal waterways at Tuross Heads (the village appears pink in the image) on the south coast of New South Wales, separated by a narrow peninsula. The Tuross River, to the south, fills a large part of the barrier estuary into which it empties (its floodplain is light green in the image). Coila Lake is a coastal lagoon with only a small freshwater input. It remains largely unfilled by sediment, and only intermittently opens to the sea after major flood events (the narrow berm that seals the mouth is unvegetated and appears white on the image).* Image: © Commonwealth of Australia, ACRES, Geoscience Australia

predominantly open, although with a series of shoals that make navigation difficult. Over the past 7000 years, the river has built over 20 kilometres into the central basin and, as the fluvial delta has prograded, alluvial floodplains have built up, primarily as a response to fluvial delta growth. Coila Lake has a much smaller catchment. The lake remains largely unfilled and generally remains closed, although it can open for short periods after large floods. Many other coastal waterways in southern New South Wales show similar characteristics (see Figure 4.12).

Mature riverine estuary/delta

Once the estuary is filled, a mature riverine estuary is developed, and the river becomes channelised as it meanders through the alluvial plains. Recognised as the ultimate stage of infill of a wave-dominated estuary, this is called a 'wave-dominated delta'. At this stage, sand is delivered directly to the adjacent coast, and the shoreline may begin to prograde.

More than two-thirds of estuarine and deltaic systems around Australia remain sediment traps, and only a few – generally the largest – rivers or creeks have reached the mature stage and supply sand directly to the adjacent coast. The estuarine plains, topped with alluvium, present rich agricultural land, extending along the floodplains that accrete upstream of the former estuarine embayment. These plains flank the larger rivers of Queensland and northern New South Wales, and have been extensively reclaimed for sugar cane, dairying and cattle grazing. Evidence of their former estuarine origin can still be detected, with low-lying flood basins that may remain below the height of highest tide level, and cut-off embayments and palaeochannels. The mouths of many of these estuaries/deltas are trained by walls and breakwaters so that they do not shoal and so that the entrance can be navigated. This can also increase the tidal volume or prism.

The Shoalhaven River in southern New South Wales has reached this final stage and can be considered a delta. The river is flanked by sandy levees, behind which are near-horizontal floodplains that have been drained for agriculture, and support dairy farming. Subtle topographic highs across the plains (usually the location of roads) are the remnants of levees that trace the course of former channels across the infilled central basin. Coring and dating of the sediments indicate that shallow alluvial topsoil overlies estuarine mud, accreted over the past 6000 years. Much of the Shoalhaven plains appear to have been infilled by 3000 years ago, at which time Swamp Sheoak (*Allocasuarina*) and

associated freshwater wetlands had replaced brackish conditions. The Shoalhaven River has contributed sand to the coast for the past few millennia, and this has enabled the development of a regressive barrier with a sequence of foredune ridges along Seven Mile Beach (described in Chapter 6).

Tide-dominated estuaries and deltas

There are more than 80 tide-dominated estuaries and 70 tide-dominated deltas along the coast of northern Australia, from the Kimberley through to Cape York Peninsula. These have received considerably less study than wave-dominated systems, and some of the most thorough investigations were undertaken in the 1970s as part of a systematic assessment of the crocodile populations. The large megatidal range in northwestern Western Australia, and macrotidal range in the Northern Territory, means that tides penetrate tens of kilometres inland along the larger systems. There are several partially infilled tidal embayments, including Darwin and Bynoe Harbours in the Northern Territory, which are equivalent to the drowned river valleys of southern Australia.

Macrotidal estuaries

The macrotidal estuaries of the Northern Territory and northern Western Australia are highly seasonal, reflecting the monsoonal climate. During the wet season, November to March, they experience heavy rains and rivers flood the extensive estuarine plains, often up to several metres water depth. At this time the river flow dominates the hydrodynamics of

TIDAL ASYMMETRY AND TIDAL BORES

Tidal asymmetry is common in macrotidal estuaries, meaning that the incoming flood tide reaches a greater velocity than the outgoing ebb tide, and that the disparity between the strength of flood and ebb tide currents increases upstream. This can result in a **tidal bore** on several of the large rivers at spring tide. A bore is a wave that moves rapidly upstream as the tide rises; this occurs on the Daly River.

Tidal asymmetry means that the incoming tide's ability to transport sediment increases with distance upstream, and this leads to development of sand shoals both in mid-channel and attached to the point bars on meanders. As a consequence, flows can become quite treacherous, and flood and ebb currents may adopt different pathways on opposite sides of the river channel. Wave action tends to be limited, although waves with sufficient fetch do shape the more exposed banks near the mouth.

the estuary and tidal influence is less prominent. Once the rains cease, the flooded plains begin to dry out and become progressively drier as the dry season advances. With reduced river flow, saltwater penetrates further and further up the estuary. The large tidal range means that tides rapidly become the dominant influence over the dry season and the system becomes well-mixed with saltwater.

Many estuaries experience a tidal range of 5–6 metres on spring tides at the mouth. Flows remain substantial – up to 2 metres per second – in the meandering sections of the estuary and develop a distinct asymmetry with greater velocities on flood tides than on the ebb tides (see box 'Tidal asymmetry and tidal bores', above). The tide can penetrate up to 100 kilometres along the larger systems such as the South and East Alligator rivers. The tide progresses up the estuary, with high tide experienced at the upstream limit several hours after it occurs at the mouth. The time at which low tide is experienced lags even further behind its occurrence at the

mouth. The tidal wave therefore deforms along the estuary and the flood tide gets progressively shorter upstream, with flood tide currents consequently accelerating relative to ebb tidal currents.

Habitats

Tidally dominated systems are strongly funnel-shaped – their mouths taper upstream as the tidal prism (the volume of water between high and low tide) decreases. Near the mouth, tidal creeks are flanked by mangroves (see Figure 4.13). The mangrove fringe is generally fairly continuous and reflects the open-water zonation that typifies much of the region. A seaward open-woodland of Apple Mangrove (*Sonneratia alba*) merges into a broader belt of Grey Mangroves (*Avicennia*), with a ground cover of River Mangrove (*Aegiceras*). The Red Mangrove (*Rhizophora stylosa*) forms a dense zone behind this, with a thicket of stilt roots. The *Rhizophora* zone is generally a zone of steep transition, commencing around mean sea level and merging into a zone of Yellow Mangrove (*Ceriops*) towards high-tide level. There are several other species that occur towards the landward margin of the mangroves, but the rear-most zone is generally dominated by the Grey Mangrove, which becomes increasingly stunted as the salinity of the substrate increases.

There is generally a landward fringe of samphire and then extensive salt flats, in places kilometres wide. Although tidal inundation is rare, these salt-encrusted areas remain unvegetated. Along much of the open coast, former shorelines may be marked by chenier ridges (described in Chapter 6), accumulations of wave-deposited sand and shell that are up to 1 to 2 metres above the coastal plain.

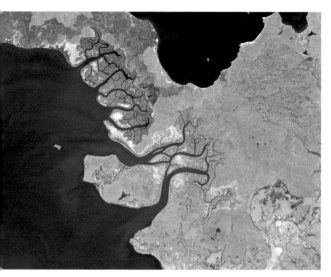

Figure 4.13 *A Landsat satellite image of the tidal creeks at the eastern end of van Diemen Gulf, in the Northern Territory. The strongly tapering channels are fringed with mangrove forests (dark green), behind which there are extensive salt flats (light pink).* Image: © Commonwealth of Australia, ACRES, Geoscience Australia

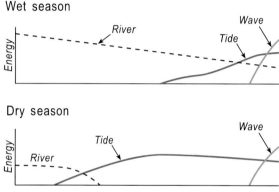

The strong tidal currents may also shape isolated sandy shoals, or linear mangrove islands within the estuarine funnel.

Morphology

The morphology typical of macrotidal estuaries is shown schematically in Figure 4.14. Three sections can be recognised: an outer estuarine funnel, a meandering central section and a straighter, narrow upstream section. Proceeding up the channel, the estuarine funnel tapers into a meandering section. The channel width continues to decrease upstream, which is a feature of tidal systems that reflects the decreasing tidal prism. The inner bank of sinuous meanders is colonised by mangroves, again with distinctive zonation of species but with shore-normal channels draining the former point bar also apparent. Outside bends are often eroding, and former meander courses may be preserved on

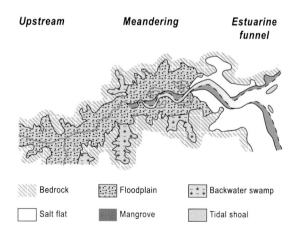

Figure 4.14 *The environments and the relative contribution of river, wave and tides along a tide-dominated estuary and the relative distribution of energy during the wet season and the dry season.*

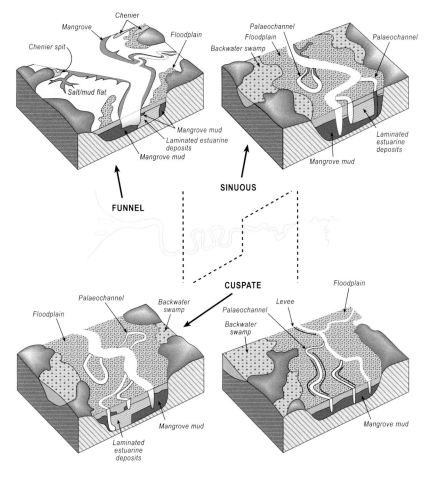

Figure 4.15 *This block diagram of a tide-dominated estuary shows the surface morphology and habitats, and the subsurface stratigraphy of estuarine funnel, sinuous meandering, cuspate meandering and upstream segments of the tidal river.*

the plains as palaeochannels (billabongs). The salt flats are replaced by freshwater sedge and grassland. These are floodplains; their sediments are composed of alluvial mud, brought down the river and deposited on the plains during floods in the wet season. The floodplains decrease in elevation away from the river, and there are low-lying paperbark wetlands (*Melaleuca* spp.) in the backwater swamps. These areas tend to remain wet throughout the year.

Sinuous meanders, like those on the Adelaide River, occur along most macrotidal rivers, but on several large estuaries, particularly the South and East Alligator rivers, there are upstream sections where the meanders are cuspate, with the inside of the meanders being pointed, and with substantial mid-channel shoals exposed in the river bed at low tide. In the upstream reaches of the tidal section of the river, the channel is much narrower and dominated by straight stretches, with angular bends, silty levees and a narrow fringe of the Crabapple Mangrove (*Sonneratia caseolaris*) and the Grey Mangrove.

Pattern of infill

The pattern of infill of tidal systems in northern and northwestern Australia differs from that of the southern wave-dominated systems. Tidal velocities entrain mud and keep it in suspension, and considerable volumes of mud can be transported into estuaries from seaward by tidal processes. Figure 4.15 shows the typical habitats and the underlying stratigraphy, as determined by extensive stratigraphic studies on the plains of the South Alligator River, and characteristic of macrotidal systems in the Top End of the Northern Territory generally.

Occasionally preserved in the eroding banks is evidence of former environments, particularly sub-fossil mangrove stumps, together with carbonate nodules containing the skeletal remains of mud lobsters and crabs. These have been radiocarbon dated to around 6000 years ago on several major river systems (South Alligator, Adelaide, Daly and Ord rivers, and King Sound) and record former extensive mangrove forests, called the 'big swamp' phase. Unlike the central basins of the wave-dominated systems, these macrotidal systems were rapidly filled with muds as sea level stabilised 6000 years ago (see Figure 4.16). The transition from former estuarine conditions has been detected by pollen analysis; there is an abundance of mangrove pollen about a metre below the plains surface, but this is replaced by grass and sedge pollen towards the surface.

The stratigraphy is shown schematically in Figure 4.15, based on the South Alligator River. Much of the plain is underlain by mangrove mud. The basal mud was deposited as sea level rose and flooded into the river valley during the early Holocene, around 8000 years ago. The 'big-swamp' phase occurred around 6000 years ago as sea level stabilised at around its present level, when mangrove forests became widespread in these embayments. The rapid accretion of sediment under the mangrove forests meant that the tide no longer penetrated into those areas most distant from the river. The mangroves were replaced by near-horizontal alluvial plains and freshwater wetlands, as found upstream at present in the monsoonal Top End. The plains have now largely transformed to be dominated by sedges and grasses, but locally with paperbark and several other tree species.

Since the replacement of mangroves, the river banks have built subtle levees, so that

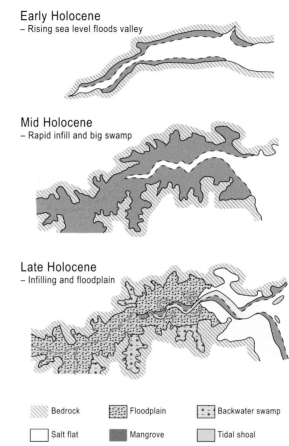

Figure 4.16 *Three stages in the evolution of a tide-dominated estuary, based on the South Alligator River. As sea level rose during the early Holocene, the prior valley was flooded and mangroves re-established over the exposed landscape. Around 6000 years ago, in mid-Holocene, tidal pumping resulted in rapid infill of the embayment with mud and a big swamp, and extensive mangroves occurred. Since that time the plains have accreted vertically and the mangroves have been replaced by saline mudflats or freshwater wetlands; the river has meandered across the plains and reworked some of the plains, and sinuous meanders have been abandoned by cut-off. The river has adopted a cuspate meandering morphology in mid-plains.*

Figure 4.17 *Extensive saline mudflats flank the Ord River at Wyndham and in the more arid parts of Western Australia.* Photo: A.D. Short

dry-season tidal flows are largely confined within the channels, whereas the monsoonal river floods the plains with freshwater. This alluvial floodplain has a convex topography typical of the monsoonal Top End; it slopes almost imperceptibly away from the levees that flank the channels, with paperbark and perennial wetlands in the backwater swamps (see Figure 4.15). By contrast, near-horizontal, bare saline flats characterise similar plains in the arid northwest of Western Australia, where there is a much smaller supply of alluvial sediment brought by the lesser floods (see Figure 4.17).

Since the development of floodplains, rivers continue actively to meander on the plains surface. The former courses of several rivers are preserved as palaeochannels (billabongs). The banks of many of the rivers are actively eroding, but there are broad depositional environments elsewhere.

Over the past 6000 years, different tide-dominated estuaries have meandered to differing extents and reworked different amounts of their floodplains. The Daly River, which has a large catchment and a considerable sand supply brought down by the strong fluvial input, appears particularly active, while the South Alligator and East Alligator rivers have meander tracts indicating similar proportions (<20 per cent) of the plains have been reworked. The exquisite meanders on the Adelaide River, however, do not appear to be actively migrating.

Some tributary tidal creeks on the Alligator rivers have undergone extension in recent years, with saltwater preferentially intruding into lower-lying ground, particularly palaeochannels that are incompletely infilled, and killing freshwater wetland vegetation. The cause of the onset of tidal creek expansion has not yet been determined unequivocally. It is believed that initial incursion coincided with feral buffalo populations, and buffalo swim channels were implicated, although more recently sea-level rise is considered as contributing. While the cause remains elusive, there have been attempts to stop the saline intrusion with the construction of barrages across creeks that appear to be threatened. There has been some success at halting the loss of wetland.

Regional characteristics of Australian estuaries

Queensland

Coastal catchments in northeast Queensland receive high rainfall and, as a result, the northern rivers are particularly sediment-laden. An average of more than 14 000 000 tonnes of sediment is discharged each year into the Great Barrier Reef province, which has significant implications for the reef (see Chapter 8). Short, steep catchments are typical of most rivers on the eastern margin of Cape York Peninsula and the Wet Tropics, because the Great Dividing Range is close to the coast. More than 4000 millimetres of rainfall is received annually at Tully, and occasional tropical cyclones also have far-reaching effects. Catchments are predominantly forested, and their near-pristine estuaries are flanked by diverse mangrove forests. Tidal range reaches its greatest amplitude (8 metres) in Broad Sound, a muddy embayment flanked by chenier ridges. Human activity has exacerbated sediment and nutrient loads in the Johnstone, Tully and Herbert Rivers. Before European settlement, the coastal plains further to the south would have supported freshwater wetlands of paperbark (*Melaleuca*), palm swamps and sedgelands, but many have since been cleared for agriculture.

The three largest Queensland rivers are the Normanby, Burdekin and Fitzroy. The Normanby drains a catchment of around 24 000 square kilometres; it flows north and empties into the southern part of Princess Charlotte Bay. The Normanby is a tidal river, with highly sinuous meanders for about 80 kilometres. A curving, 70-kilometre-long series of ridges, forming distinct cheniers, overlie the muddy salt flats at its mouth, and these coalesce into a series of beach ridges to the more exposed northeast.

The larger Burdekin and Fitzroy catchments have undergone more widespread changes of land use. Across the Burdekin River catchment, sugar cane and cattle grazing occupy much of the extensive plains flanking the rivers. Prawn aquaculture has been established on the tidal flats. Damming and

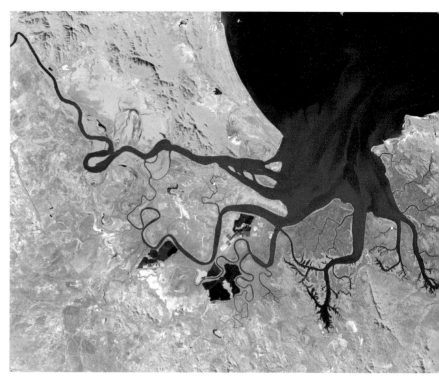

Figure 4.18 *A Landsat satellite image of the Fitzroy River mouth, Queensland. Tidal channels of several different styles can be seen, including dense dendritic creeks, tapering macrotidal channels and meander palaeochannels.* Image: © Commonwealth of Australia, ACRES, Geoscience Australia

other land-use issues have led to replacement of sedge species with invasive weeds.

The Fitzroy River has the biggest catchment of any coastal river around Australia (it is second only to the Murray–Darling river system) and has highly variable flow. Much of the 150 000 square-kilometre catchment supports grazing, as well as mining activities. The lower estuary is always turbid, due to strong currents associated with a tidal range of up to 4 metres, and there are extensive mangrove forests fringing the tidal reach (see Figure 4.18).

Southeast Queensland contrasts with the north of the state in several respects. First, whereas the energy along the coast of northeast Queensland is reduced by the Great Barrier Reef, southeast Queensland is exposed to incoming swell. It has received abundant sand carried north by longshore drift, forming large sand islands (see Chapter 6), behind which there are extensive embayments. It is also highly urbanised and has a rapidly growing population. Hervey Bay is partially enclosed behind Fraser and Bribie islands, and Moreton Bay is sheltered by Moreton and North Stradbroke islands. The bays open to the north, and contain shallow waters, sandy seafloor and stands of mangroves along the shore. The salt flats, saltmarshes, mudbanks and seagrass-covered sandbanks support dugong and turtle populations. Hervey Bay is seasonally important for the whales that visit during their annual migration along the coast.

New South Wales

Along the coast of New South Wales, estuaries vary primarily as a consequence of their physical setting, but also as a result of the degree of anthropogenic modification. The north coast has broad estuarine plains that have been intensively settled; the Sydney Basin is dominated by drowned river valleys that cut through the sandstone plateau; and the southern coast is composed of smaller catchments and an embayed coast.

Estuarine plains of northern New South Wales

A series of rivers in northern New South Wales have infilled their lower sections and are mature riverine estuaries that have reached the wave-dominated delta stage. The Tweed River, which marks the state boundary, has training walls at its mouth to maintain a navigable entrance. These walls interrupted sand movement up the coast, necessitating construction of a major sand-pumping operation (see page 20). The Richmond River at Ballina, the Clarence River at Yamba–Iluka, the Macleay River at South West Rocks and the Hastings River at Port Macquarie, as well as some of the smaller rivers, also have training walls at their mouth (see Figure 4.19). The Nambucca, Bellinger, Hastings, Macleay and Manning rivers, on the New South Wales mid-north coast, each has a bar at their entrance.

The floodplains of the Richmond, Clarence, Macleay and Hastings rivers have been converted to grazing and sugar cane. Port Macquarie, on the Hastings River, was one of the earliest centres of sugar cane production dating back to 1823. Since then, oysters and prawns have also become important on many of these waterways. The Clarence has a large fishery (mullet, mud crab, bream and prawns). Its plains are inundated during big floods, with the lower estuary comprising a number of palaeochannels that also carry water. Newcastle is a major port city on the banks of the Hunter River, and Port Stephens is another sheltered harbour that is interconnected by the Myall River to Myall Lakes, the northern of which are mainly freshwater (see Figure 6.29). Wallis Lake at Forster has about 20 per cent of the seagrass in New South Wales. Acid sulphate soils result from the oxidation of sulphides in the estuarine muds that underlie the plains following drainage for agriculture, and have resulted in periodic fish kills in several of these systems.

Drowned river valleys of Sydney

The massive Permian–Triassic Hawkesbury sandstone in the Sydney Basin region has been denuded into a series of deeply incised, drowned river valleys, such as Broken Bay, Sydney Harbour, Botany Bay and Port Hacking. Particularly spectacular is the mouth of the Hawkesbury River, which rises to the southwest of Sydney (where it is known as the Nepean River). It runs in steep-sided, deep valleys clothed in eucalypt forest, and exits at Broken Bay (see Figure 4.8). The estuary extends 50 kilometres inland to the tidal limit. The river has brought down sufficient sediment from its catchment for an extensive fluvial delta to build along the incised valley. Various tributary creeks also incised into the sandstone meet the main estuary, and these are filled in as a function of their catchment size. On the southern side of Sydney, Port Hacking is a more compressed version of a similar drowned river valley.

Embayed coast of southern New South Wales

The coast between Wollongong and Jervis Bay is an embayed coast comprising sandy beaches flanked by rocky headlands. Individual coastal compartments are closely linked with the catchments of different rivers and streams, and the watersheds between catchments are expressed on the coast as headlands. Lake Illawarra, which occupies a broad valley, is an example of a barrier estuary. Subsurface seismic studies across the lake indicate that the former courses of the two coalescing palaeochannels draining to the east – the former courses of the two creeks that feed into the lake, Mullet Creek and Macquarie Rivulet (see Figure 4.6) – incised into the underlying Pleistocene surface. The Minnamurra

River drains off the steep Illawarra escarpment and has undergone a transition from a wave-dominated estuary with central mud basin to an infilled, mature riverine estuary. Its central basin, an area called Terragong Swamp, was infilled and vegetated with wetlands that were drained in the early 20th century and now support dairy farming.

The Shoalhaven River is an infilled system, as described above, and has contributed sand that is now deposited in the foredune ridges that parallel the adjacent Seven Mile Beach. Whether the river still provides sand to the coast is uncertain, because there have been two major human impacts.

First, the river used to exit to the sea at Shoalhaven Heads, but when the shoals at its mouth proved treacherous to navigate, an early settler and local landowner in the region, Alexander Berry, had his

Figure 4.19 Many wave-dominated deltas have formed in northern New South Wales. The plains have been used intensively, both for settlement and agriculture, and the mouths of large rivers are controlled by training walls to make them navigable, as shown here at the mouth of the Brunswick River. Photo: A.D. Short

Figure 4.20 Pambula River, southern New South Wales, a wave-dominated estuary with a central basin of 1.5 square kilometres. Photo: A.D. Short

by the successive stages that can be seen in the adjacent estuarine systems. The far-south coast of New South Wales consists of a series of small estuarine systems. Many are coastal lagoons that are intermittently open to the sea. A few larger barrier estuaries include Merimbula Lake and the Wagonga River at Narooma, and the drowned Pambula River mouth (see Figure 4.20).

Victoria

The east coast of Victoria contains similar wave-dominated estuaries as in New South Wales. The mouth of the Snowy River, at Marlo, does not reflect the mighty river that rises on Mount Kosciuszko, since diversion of a major part of its waters to the Murray and Murrumbidgee rivers was a part of the Snowy Mountains Hydro-Electric Scheme in the mid-20th century. By contrast, Gippsland Lakes represent one of the longer coastal lagoons, which since 1889 have been connected directly to the sea at Lakes Entrance, resulting in saltwater intruding into the former freshwater lakes. Corner Inlet is another unusual embayment: sheltered behind Wilsons Promontory, it contains broad tidal flats and is the site of the southernmost mangroves in the world. Port Phillip Bay is a large embayment with an area of 1950 square kilometres; it has a deep mouth through which there are strong tidal currents, but the shallower southern part of the bay is being dredged to enable large vessels to gain access to the Port of Melbourne. Westernport Bay occupies a tectonic depression and its mouth is partially occluded by Phillip and French islands. Along the high-energy west coast, most of the estuarine mouths are diverted or sealed by the waves.

convict labourers construct a 200-metre-long artificial canal. Now known as Berry's Canal, it continues to widen, and directs the flow of the Shoalhaven River to exit at Crookhaven Heads. However, following major floods, the former mouth at Shoalhaven Heads often opens temporarily. Second, flash flooding, a feature of the Shoalhaven River, has been considerably smoothed as a result of construction of the Tallowa dam, upstream of Nowra.

Jervis Bay is a large embayment formed in a geological depression. It contrasts with the smaller estuarine and coastal lagoon systems elsewhere on the New South Wales coast, because it is not sealed across its entrance with a sand barrier. In this respect the bay is more similar to the drowned river valleys. The topographic control of the large bedrock-flanked embayment, and limited sediment supply from the small catchments that feed into it, mean that Jervis Bay has not undergone the history of infill that is recorded

Tasmania

The Tasmanian coast contains a wide range of estuarine environments. The mesotidal north coast experiences limited wave action in the west, with several tide-dominated estuaries in the lee of Robbins Island and at Stanley. These grade to generally small, wave-dominated barrier estuaries as wave energy increases to the east, with the central Port Sorell and Tamar River both larger, elongated drowned river valleys.

The east coast is dominated by generally small barrier estuaries and some saline coastal lagoons, together with occasional, larger drowned river valleys including Georges Bay, Little Swanport, Prosser Bay, Blackmans Bay and Port Arthur. The highly indented southeast coast has a wide range of generally sediment-deficient drowned river valleys and bays, with the Derwent being the only sizable river entering a drowned river valley.

The west coast has several large pristine drowned river valleys, including the large Port Davey and Macquarie Harbour, and the sinuous Pieman River. Numerous small rivers and creeks flow into usually exposed, small blocked river mouths, as in the case of the Henty River (see Figure 4.21).

South Australia

There are few rivers along the relatively arid coasts of South Australia and Western Australia. Although the mouth of the Murray–Darling basin is found here, the river rises in the eastern highlands and is no longer carrying the flows that it used to. In addition, the mouth has been intermittently closed. Despite a length of over 2500 kilometres, the river empties

Figure 4.21 The deflected mouth of the Henty River, a wave-dominated estuary in western Tasmania. Photo: A.D. Short

into shallow Lake Alexandrina and then divides into five channels and flows around Hindmarsh Island before making an unremarkable exit into Encounter Bay at the small Murray Mouth opening (see Figure 4.3). At times of low flow, the river fails completely to reach the sea. In fact, in historical times the river's mouth became choked with sand. Barrages set in place since the 1940s now prevent saline water penetrating up the Murray system. This means that brackish fish are no longer found upstream, either. Some of the water enters the Coorong, a 2–3-kilometre-wide coastal lagoon system stretching for at least 100 kilometres from the river mouth and separated from the sea by Younghusband Peninsula, a sparsely vegetated barrier covered by transgressive dunes. The Coorong, although hypersaline (with salinity up to twice that of seawater), supports a rich estuarine ecosystem.

Spencer Gulf and St Vincent Gulf are two large embayments that are examples of inverse estuaries with hypersaline waters at their upper end. These are relatively sheltered from wave action, and contain rich seagrass beds and have fringing intertidal communities, including the mangrove *Avicennia* and saltmarsh. Between Davenport Creek near Ceduna and Poison Creek east of Esperance, a 1150-kilometre section of the Great Australian Bight has not a single creek or stream reaching the coast.

Western Australia

Estuaries around the more humid southwest coast of Western Australia occur in an environment of very limited tidal action, highly episodic and seasonal river flow, and an energetic wave climate that builds barriers across the mouths and blocks inlets. The barrier estuaries include many that are generally larger, permanently open systems, such as the Oyster, Nornalup and Swan. These contrast with those that are seasonally open along the south coast, including the Broke, Irwin and Parry systems, as well as Wilson Inlet. There is also a series of smaller, normally closed coastal lagoons including the Beaufort, Gordon, Hamersley and Stokes. Drowned river valley estuaries also occur around the coast, with most normally closed or seasonally open, like the Warren, Margaret, Buller and Hutt.

The northern half of Western Australia is particularly arid, and large rivers such as the Gascoyne flow only intermittently. Shark Bay is a large embayment on the central Western Australia coast, and although not strictly an estuary as it does not receive river input, it contains unique hypersaline environments in which stromatolites are presently forming. The Carnarvon–Pilbara coast has a series of larger, usually dry rivers that drain the Pilbara region, most only flowing after heavy rain

associated with occasional tropical cyclones. Their mouths generally consist of sequences of low beach ridges. These include the Wooramel, Gascoyne, Ashburton, Cane, Robe, Maitland, Harding, Sherlock, Peewah, Yule, Turner and De Grey (the Maitland River delta is discussed in Chapter 6). The inlets are dominated by tidal flow, are lined with extensive mangrove forests and backed by wide salt flats. The distribution of mangroves varies across these systems, responding to geomorphologically defined habitats that reflect gradients in the key environmental variables such as salinity, ground water flux, temperature and water level.

King Sound has Australia's largest tidal range, up to 12 metres at springs. The Fitzroy River, which flows into the southern end of the Sound, is highly seasonal, but the system is flushed by the large tides. Mudflats are undergoing erosion in many places. The remains of mangrove stumps that grew 6000 years ago can be seen at several places. Cambridge Gulf, into which the Ord River flows, represents another large embayment with a tide range up to 9 metres. Since the damming of the Ord River in 1972, the lower Gulf has continued to infill with muddy sediments derived from seawards.

Northern Territory

The Top End coast of the Northern Territory is highly monsoonal, with about 1600 millimetres of rain falling at Darwin and along the southern shore of van Diemen Gulf. The Daly River has the largest catchment, and the macrotidal estuarine section has a series of large meanders that are mirrored in the scroll bars of palaeochannels covering much of the plains. Darwin Harbour is a broad embayment with

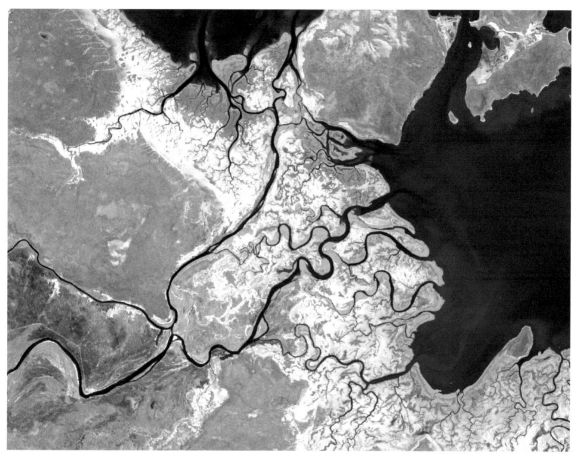

Figure 4.22 *A Landsat satellite image of the delta of the McArthur River. Although this river mouth has a number of channels like the distributaries of a more complex delta, there are only two that carry most of the flow: the westernmost channel (Carrington Channel), which carries most of the flow from Batten Creek, and the main McArthur River channel, which meanders across the delta and maintains a near-constant channel width.* Image: © Commonwealth of Australia, ACRES, Geoscience Australia

a series of prominent arms that have not infilled. It contains one of the largest stands of mangrove forests in northern Australia.

Several large rivers empty into van Diemen Gulf in the wet season. The Adelaide River contains a series of very symmetrical meanders with mangrove forest on their inner banks. The South and East Alligator rivers each have prominent cuspate meandering sections upstream of the sinuous meandering section (see Figure 4.15).

Smaller systems such as the Wildman and West Alligator rivers, and the series of creeks at the eastern end of the gulf (see Figure 4.13) contain fine stands of zoned mangrove lining their estuarine funnel. The Mary River, however, is an exception; plains that have been shown as having evolved through the infill of the prior estuarine embayment have been largely cut off from the sea, until saltwater intruded into the plains via two tidal creeks (see Chapter 9).

Similar mangrove-lined systems occur along the northern coast of Arnhem Land and around the Gulf of Carpentaria. Many of these are highly inaccessible, such as the near-pristine freshwater Arafura Swamp, which appears also to have formed over estuarine wetlands in a similar way to those systems shown in Figure 4.16.

The only large river on the west coast of the Gulf of Carpentaria is the Roper River. Along the southern shore there are several rivers. The McArthur River enters opposite the Sir Edward Pellew Islands and forms a delta with many bifurcating channels (see Figure 4.22). This is because Batten Creek flows parallel to the McArthur River itself and there is some exchange of water at the apex of the delta. However, most of the flow of Batten Creek continues down Carrington Channel, whereas the McArthur River meanders towards Centre Island. There are several distributaries that are strongly tapering, such that it is only possible to navigate from the main river into these creeks at high tide. The river itself is broad and maintains a channel of near uniform width, and it appears that the distributaries mark former courses of the McArthur River that have infilled to become tide-dominated after the main river channel adopted a new location (see Figure 4.22).

Several other rivers flow into the southern shore of the Gulf of Carpentaria, which is a very low wave-energy and low tide-range environment. The Nicholson, Albert, Leichhardt and Norman rivers deliver large volumes of mud and have built wide, low-gradient floodplains that grade into intertidal salt flats fringed by cheniers and mangroves. On the eastern margin of the Gulf of Carpentaria, there are several rivers that have sinuous channels crossing alluvial fan deltas. These are subject to sand blockage at their mouths. They deliver sand and mud to the shore, building long, sandy beaches fronted by extensive intertidal mud flats. The Calvert River (see Figure 4.23) is typical of several others, such as the Gilbert, Mitchell and Archer rivers, which drain the low, seasonally wet Cape York Peninsula.

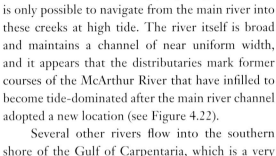

Conclusion

The great variety of estuaries and deltas are important components of the Australian coast. They are natural centres of biodiversity and productivity, and many have become the focus of ports, cities and towns, transportation and trade, as well as recreation and tourism. The Australian continent is generally of low elevation, and much of it receives little rainfall.

Figure 4.23 The Calvert River, located in the southern Gulf of Carpentaria, shows a wave-dominated system. The effect of shoaling waves in building sand across the mouth of the river is clearly visible.
Photo: A.D. Short

As a consequence, rivers play only a minor role in shaping the landforms at their mouths.

The two factors that are particularly important are waves and tides. Where tides are dominant, coasts are generally muddy and the tidal channel is usually funnel-shaped, broad at the mouth but tapering rapidly upstream where it is characteristically meandering. Across northern Australia tidal channels are fringed with extensive mangrove forests and backed by salt flats. Where waves dominate, the coast is typically sandy, and waves align shoals and barriers parallel to the coast. Wave-built bars may close the entrance to waterways, unless river flow is sufficient to keep the mouth open or to re-open the inlet.

As we emphasised in Chapter 2, southern Australia is exposed to high-energy Southern Ocean swell and its coasts are predominantly sandy, whereas the largest tides occur in northern Australia, and its estuaries and parts of the open coast are predominantly muddy. Consequently, there is a preponderance of wave-dominated estuaries in southern Australia, especially along the southeastern and southwestern seaboards, where the numerous adjacent catchments terminate in a diversity of estuaries, coastal lagoons and drowned river valleys. In northern Australia, particularly along the macrotidal coast of the Northern Territory and the megatidal north of Western Australia, there are many tide-dominated estuaries, characterised by flooding in the wet season, but becoming increasingly saline during the dry season.

All estuaries are infilling with sediment, and a significant threshold occurs when the estuary becomes a delta, changing from predominantly retaining sediment within it, to a by-passing system which supplies river sand to the coast. Meanwhile, at the intermediate stages of infill, estuaries contain a rich diversity of habitats as well as attractive and productive freshwater and intertidal wetlands, which make them a magnet for people, both as places to live and to visit for recreation.

BEACHES

Introduction

THE AUSTRALIAN COASTLINE has an impressive 10 685 beaches. In fact, beaches make up half of the open coast. Amazingly, every single one of these beaches can be classified into one of 15 distinctive beach types, based on the nature of their waves, tides and sediment.

Most Australian beaches are bordered by some natural structure such as a headland, rocks, reef or inlet. They are the most obvious result of the marine and atmospheric processes acting on Australia's coastal geology. Australia's beaches consist predominantly of fine to medium, clean white to yellow sand, with only a relatively few gravel, cobble and boulder beaches. Beaches form in environments ranging from low waves and high tides in the north, to some of the world's highest-energy beaches in the south.

Definitions

A beach is a wave-deposited accumulation of sediment – usually sand, but ranging in size up to boulders – deposited between the upper swash limit and wave base. Most beaches contain three or four zones:

1. the inner swash zone, which extends from the upper limit of swash to the shoreline and includes the upper-horizontal, usually dry, berm and the sloping beachface, where waves run up and down the beach as swash

2. many beaches also contain a surf zone, extending from the shoreline to where the waves break (the breaker zone)
3. seaward of the surf zone is the nearshore zone, extending from the breaker zone to wave base
4. wave base is the depth at which waves begin to interact with the seabed and can transport sand toward the beach, and out to which sand is transported during big seas and accompanying beach erosion.

Waves and currents recycle sand from the wave base, across the entire beach system (that is, the swash, surf and nearshore zones). The depth of wave base and the width of the beach system increases with wave height, reaching as much as 30 metres deep and 2–3 kilometres wide along the high-energy southern coast, while on very low-energy beaches it only extends to low tide.

Dynamic zones

Every beach has three dynamic zones related to the transformation, breaking and final demise of ocean waves (see Figure 5.1). Deepwater ocean waves are called progressive waves where the wave form progresses forward while the sea surface simply moves up and down. Progressive waves begin to 'feel' and interact with the seabed as they enter the nearshore zone, in a process called 'wave shoaling'. Here, beginning at wave base, they increasingly expend energy on the seabed, forming ripples and generally transporting sand landward.

The nearshore zone is always a concave, upward-sloping seabed, with a gradient between 0.8 and 1 degree, and covered in small, wave-formed sand ripples. It is usually the largest zone of the beach, extending tens to hundreds of metres seaward of the shoreline. The nearshore zone extends shoreward

AUSTRALIA'S SHORTEST AND LONGEST BEACHES

Australian beaches have a relatively short average length of 1.37 kilometres. Most states have no beaches longer than 26 kilometres, and there are 22 beaches only 20 metres in length. However, there are a very few relatively long beaches (see Table 5.1). The three longest beaches – the Coorong (South Australia), Ninety Mile (Victoria) and Eighty Mile (Western Australia) – exceed 200 kilometres in length and are all located within sedimentary basins, devoid of bedrock. Seventy-Five Mile beach (89 kilometres) is part of Fraser Island (Queensland).

[1] The world's longest beach is possibly the 390-kilometre-long sandy coast of Rio Grande do Sul, in southern Brazil.
[2] Continuous beach with no interruptions.
[3] Beach is interrupted by rocks, reefs and/or creeks.
[4] See section on Australian beach types for full description.

	continuous[2] km	non-continuous[3] km	beach type[4]
Coorong– Middleton, SA	194	212	wave-dominated
Ninety Mile, Vic.	125	222	wave-dominated
Seventy-Five Mile (Fraser Is.), Qld	89	89	wave-dominated
Eighty Mile, WA	73	222	tide-modified
Ten Mile, NSW	26	26	wave-dominated
Foochow, (Flinders Is.), Tas.	24	24	wave-dominated
Dooley Point (N), NT	20	20	tide-modified

Table 5.1 Australia's longest beaches[1]

to the breaker or surf zone, where waves break and transform from progressive to translatory waves, in which the water physically moves, or is translated, shoreward as a broken wave and white water. This is the zone where the potential energy contained in the progressive waves is released as kinetic energy. As the broken wave moves shoreward across the surf zone, it generates a range of surf zone currents. These currents can move water and sediment shoreward, longshore and seaward, and in the process shape the sandy seabed into bars, channels and troughs.

Bars are shallower accumulations of sand over which waves break, while the troughs are deeper, shore-parallel sections between the bar and beach. Channels are deeper sections running seaward through the bar. As soon as the broken wave, white water or wave bore reaches the shoreline, it immediately collapses and runs up the beach as swash, within the swash zone. Swash is a thin flow of water that runs up the steeper beachface to an elevation 2 to 3 metres above the water level, until gravity and percolation into the sand cause it to stop and run back down the beach as backwash. The slope of the beachface increases with the size of the sediment, and ranges from as low as 1 degree with fine sand, to 4–8 degrees with medium sand and up to 15–20 degrees with cobbles and boulders (see Figure 5.2).

In southern Australia, where tides are generally low, the three zones of shoaling, surf and swash are relatively stationary, occupying the same location at high and low tides. However, in northern Australia, where tides are higher and waves lower, these zones can shift considerable distances with the rise and fall of the tide, producing a wide intertidal zone, in places hundreds of metres wide. The fact that these

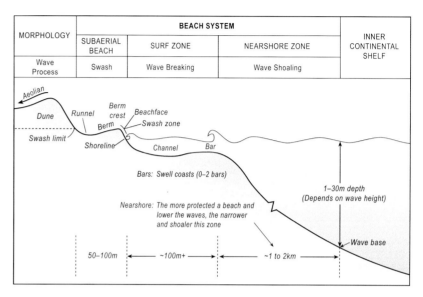

Figure 5.1 *An idealised cross-section of a wave-dominated beach system, showing the swash zone, which contains the subaerial or 'dry' beach (runnel, berm and beachface) dominated by swash; the energetic surf zone (bars and channels) with its breaking waves; and the nearshore zone extending out to wave base, where waves shoal to build a concave upward slope.*

zones are stationary in the south and mobile in the north has a profound impact on the nature of the associated beach systems, as we discuss below.

Beach sediments

All beaches consist of some form of sediment, usually either land-derived (terrestrial or terrigenous) quartz, or marine-derived carbonate material (see Chapter 1). Sediment can be classified according to size, mineralogy and colour. The vast majority of Australian beaches are composed of sand, with only a small percentage (<1 per cent) made up of cobbles, and very few of boulders. The sand on Australia's wave-dominated beaches is predominantly fine to medium (0.2–0.4 millimetre), increasing to cobble size on lower-energy, tide-modified beaches (3–5 millimetres) and particularly the very low-energy, tide-dominated beaches (4–8 millimetres), where much of the coarser material is composed of shell and coral debris. Their mineralogy consists of either quartz grains or calcium carbonate in the form of shell, algae and coral debris. Some regions have small percentages of dark, heavy minerals. Both quartz and carbonate sediments tend to be white through to yellow and orange in colour.

Australian beach types

Although beaches are a product of waves and sand, the tides also play a role in beach dynamics. The interaction between beach sediments, waves and tides determines the beach type, and variations in the combination of these result in three major Australian beach systems: wave-dominated, tide-modified and tide-dominated, which contain a total of 13 beach types. If we then include the nature of the shoreline, two more beach types occur, bringing the total number to 15 beach types. Figure 5.3 plots the general distribution of each beach type around the coast, and shows the dominance of wave-dominated beaches across the south, and tide-modified and tide-dominated beaches across the north.

Figure 5.2 a) *A steep, cobble-boulder beach at Catfish Bay, Western Australia;* *(b)* *A low-gradient, fine-sand beach at Ball Bay, Queensland.* Photos: A.D. Short

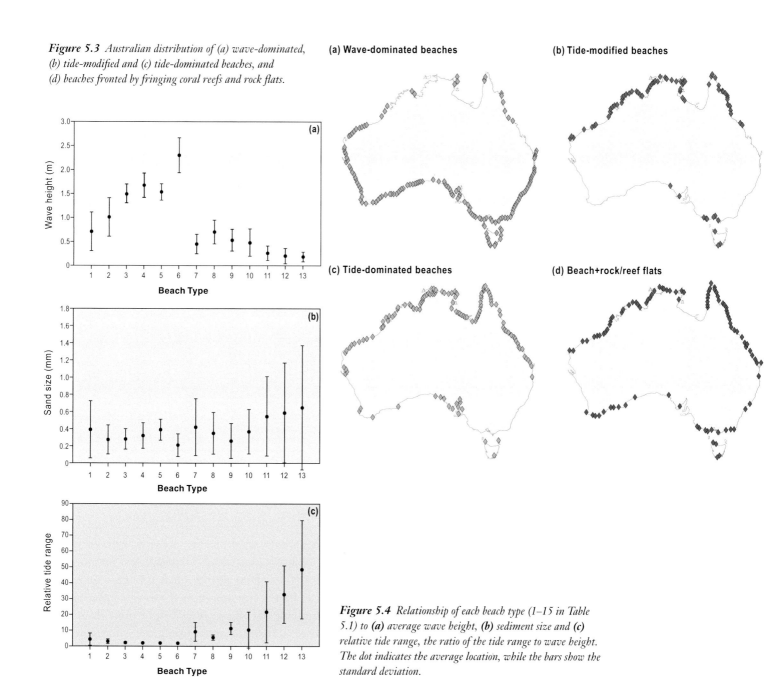

Figure 5.3 *Australian distribution of (a) wave-dominated, (b) tide-modified and (c) tide-dominated beaches, and (d) beaches fronted by fringing coral reefs and rock flats.*

(a) Wave-dominated beaches

(b) Tide-modified beaches

(c) Tide-dominated beaches

(d) Beach+rock/reef flats

Figure 5.4 *Relationship of each beach type (1–15 in Table 5.1) to (a) average wave height, (b) sediment size and (c) relative tide range, the ratio of the tide range to wave height. The dot indicates the average location, while the bars show the standard deviation.*

111

Beach type[1]	(a) Number of beaches									(b) Beach length (km)								
	Qld	NSW	Vic	Tas	SA	WA	NT	Total	%	Qld	NSW	Vic	Tas	SA	WA	NT	Total	%
Wave-dominated																		
1 Dissipative	0	0	0	0	14	4	0	18	0.2	0	0	0	0	270	8	0	278	1.9
2 Longshore bar & trough	0	0	0	0	7	1	0	8	0.1	0	0	0	0	15	3	0	18	0.1
3 Rhythmic bar & beach	7	73	83	2	33	81	0	279	2.6	112	183	280	0	91	134	0	800	5.4
4 Transverse bar & rip	34	237	170	193	159	221	14	1028	9.6	241	525	353	259	288	660	45	2370	16.1
5 Low tide terrace	13	195	116	174	164	241	28	931	8.7	97	200	125	135	203	595	50	1405	9.6
6 Reflective	40	216	153	786	571	843	61	2670	25	75	67	73	407	354	1053	34	2062	14.0
Tide-modified																		
(R=reflective)																		
7 R+low tide terrace	478	0	32	28	42	60	114	754	7.1	792	0	19	17	45	50	182	1105	7.5
8 R+low tide rips	127	0	0	1	2	8	41	179	1.7	570	0	0	0	16	10	61	658	4.5
9 Ultradissipative	95	0	0	8	0	66	11	180	1.7	201	0	0	16	0	350	58	625	4.3
Tide-dominated																		
10 R+sand ridges	195	0	0	33	29	35	98	390	3.7	482	0	0	18	44	38	300	887	6.0
11 R+sand flats	485	0	72	35	364	743	437	2136	20.0	751	0	119	22	517	749	503	2641	18.0
12 R+tidal flats	87	0	0	2	69	552	127	837	7.8	111	0	0	3	177	411	141	843	5.7
13 R+tidal flats (mud)	12	0	0	0	0	110	120	242	2.3	57	0	0	0	0	54	172	283	1.9
Rock & reef flats																		
14 R+rock flats	57	0	66	7	0	247	402	779	7.3	24	0	20	1	0	192	314	551	3.3
15 R+coral reef	20	0	0	0	0	199	35	254	2.4	13	0	0	0	0	92	36	142	1.0
Total	1650	721	692	1269	1454	3411	1488	10685	100	3525	974	989	878	2020	4398	1902	14686	100.0
mean km										2.14	1.35	1.43	0.69	1.39	1.29	1.28	1.37	

Table 5.2 Australian beach types by state, (a) number of beaches and (b) cumulative beach length (km)

[1] See Figures 5.6, 5.17, 5.21 and 5.26 for definition sketches of beach types 1–15.

Distribution of beach types

Table 5.2 lists the number and length of each beach type for each state and territory. Beach types 1–6 are wave-dominated, 7–9 are tide modified, 10–13 are tide-dominated, and beaches 14 and 15 are influenced by fringing rock or reef flats. Figure 5.4 plots the relationship of each beach type to wave height, sand size and relative tidal range. Australia's wave-dominated beaches occur in areas of higher average waves (0.5–2.5 metres), are composed of fine to medium sand and have a relative tide range ($RTR=TR/H_b$) less than 3, usually less than 1, meaning the tide range (TR) is no more than 1 to 3 times the average wave height (H_b). In Figure 5.5a the wave-dominated beaches are further classified into higher-energy dissipative, moderate to high-energy intermediate and lower-energy reflective, based on the size of the waves and sediment. Table 5.2 provides the average range of beach types under average wave conditions. It should be noted that the wave-dominated types can change as waves increase, shifting to a higher-energy type, or they can decrease, shifting gradually to a lower-energy type.

Tide-modified beaches in Australia have lower waves (0.5–1 metre) and coarser sand (0.3–0.5 millimetre), and the tidal range exceeds the wave height by a factor of 3 to 10. Tide-dominated beaches have still lower waves (<0.5 metre) and coarser sand (>0.5 millimetre), and the tide reaches from 10 to 50 times the wave height. Figure 5.5b illustrates the relationship of the tide-modified and tide-dominated beaches to wave height, period and sand size. To use the chart, determine the breaker wave height, period (T) and grain size (mm). Read off the wave height and sand size, then use the period to determine where the boundary of reflective/intermediate, or intermediate/dissipative beaches lies. $\Omega = 1$ along solid T lines and 6 along dashed T lines. Below the solid lines $\Omega < 1$ and the beach is reflective, above the dashed lines $\Omega > 6$ and the beach is dissipative, between the solid and dashed lines Ω is between 1 and 6 and the beach is intermediate. ($\Omega = Hb/WsT$ and is known as the dimensionless fall velocity.) The mean positions of 13 Australian beach types are located on Figure 5.4.

As wave conditions change a beach can move from one type to another. In Figure 5.5b, the tide-modified and tide-dominated beaches are all located in areas exposed to both low waves (<0.5 metre) and short period waves (3–5 seconds), thereby reducing the level of wave height required for the reflective, intermediate and dissipative thresholds, respectively.

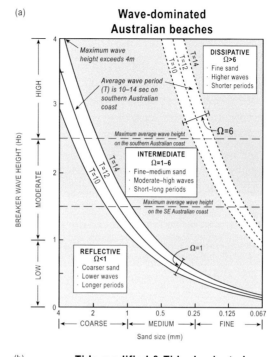

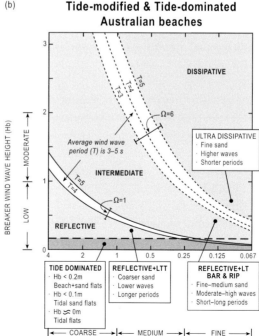

Figure 5.5 *A plot of breaker wave height versus sand size, together with wave period, that can be used to determine the approximate beach type for (**a**) the wave-dominated southern half of Australia and (**b**) the tide-modified and tide-dominated northern half.*

Figures 5.4 and 5.5 indicate that subtle variations in the combination of wave height, tidal range and sand size have a profound impact on the type of beaches around the Australian coast. While to the untrained eye all beaches may look the same – an area of bare sand and breaking waves – there are major differences between beaches caused by subtle variations in the above parameters, which in turn have an impact on the size, shape and behaviour of each beach system, as well as the longer-term evolution of the shoreline. All of these beach systems are discussed in further detail in the following sections, and their longer-term evolution is examined in Chapter 6.

Wave-dominated beaches

Distribution

Wave-dominated beaches predominate around the southern half of the Australian continent. Beaches are classified as wave-dominated when the spring tidal range is less than three times the average breaker wave height. The only parts of northern Australia with wave-dominated beaches are those on eastern Arnhem Land and eastern Cape York Peninsula that are exposed to strong trade winds and moderate wind waves averaging up to 1 metre high and where tides are relatively low (~ 2–3 metres). The following description of each wave-dominated beach type begins with the highest energy (dissipative) down to the lowest energy (reflective).

Dissipative

Dissipative (D) beaches (see Figure 5.6) form in areas of persistent high waves (H_b>2–3 metres) on beaches composed of fine sand (~0.2 millimetres).

On the swell-dominated coast of Australia, where waves are high, wave periods are also long (10–14 seconds), which means very high waves are required along with the fine sand to produce this beach type. For this reason there are only 18 fully dissipative beaches in Australia (see Table 5.2), all facing into the Southern Ocean and all located on exposed sections of the South Australian and Western Australian coast. They do not occur on the east or west coast, except during storms, as waves are usually not large enough, nor do they occur in northern Australia – where the waves are just too small.

The combination of high waves and fine sand maintains a wide, low-gradient surf zone up to 500 metres wide with usually two to occasionally three shore-parallel bars that are separated by subdued troughs. The beachface is composed of fine sand, and is always wide, low and firm – firm enough to support the weight of a vehicle. The popular Goolwa Beach, south of Adelaide, is one of the more accessible dissipative beaches, and on summer weekends is often covered with vehicles.

Wave breaking begins several hundred metres offshore on the outer bar, as high (2–3 metres) spilling breakers; that is, waves that break over a gently rising slope and cause the crest to spill down the wave face. The waves reform in the outer trough to break again and again on the inner bar or bars (see Figure 5.7). In this way they dissipate their energy across the wide surf zone, which is the origin of the name 'dissipative'. In the process of continual breaking and re-breaking across the surf zone, the individual waves decrease in height and may be indiscernible at the shoreline. The energy and water that commenced breaking in the original wave is gradually transferred in crossing the surf zone to a lower frequency movement of water, called a standing wave.

At the shoreline, the standing wave is manifest as a periodic (every 60–120 seconds) rise in the water level (set-up), followed by a rapid fall in the water level (set-down). As a rule of thumb, the height of the set-up is 0.3–0.5 times the height of the breaking waves (that is, 1–1.5 metres for a 3-metre wave). Because the wave is standing, the water moves with the wave in a shoreward direction during set-up, and a seaward direction during set-down, with velocities between 1 to 2 metres per second closer to the seabed. As the water continues to set-down, the next wave is building up in the inner surf zone – often as a substantial wave bore – up to 1 metre and higher. The bore then flows across the wide, low-gradient swash zone and continues to rise as more water moves shoreward in the set-up.

The beach is planed down to a wide, low-gradient swash zone, because of the fine sand and the nature of the large, low-frequency standing waves, with the high tide swash reaching to the back of the beach, often leaving no dry sand at high tide. Dissipative beaches tend to be relatively straight and uniform alongshore, with either no rip currents or a few widely spaced rips.

Longshore bar and trough

Longshore bar and trough (LBT) beaches are characterised by waves averaging 1.5–2 metres in height. These break over a near-continuous, longshore bar located between 100–150 metres seaward of the beach, separated from the beach by a 50–100-metre wide, 2–3-metre deep longshore trough (see Figures 5.6 and 5.8). Sand on these beaches averages 0.4 millimetres in size, significantly coarser than the dissipative beach sand, and this

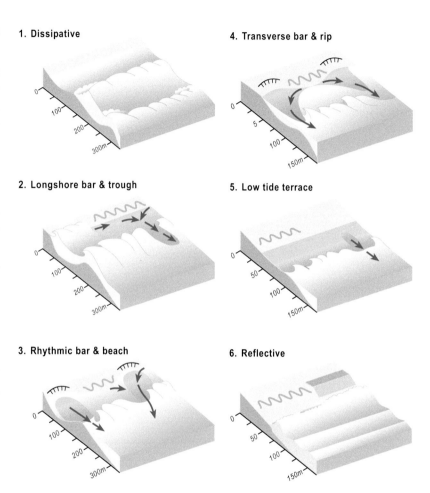

Figure 5.6 *The six wave-dominated beach types, ranging from the higher-energy dissipative beaches (1), through the intermediate rip-dominated beaches (2–5) and the reflective beach (6). Arrows indicate rip currents, curving orange beach lines are beach cusps and black curving lines indicate erosion scarps. During changing wave conditions a beach can shift from one type to another.*

results in steeper gradients on the bars and beach, and a deeper trough. The beachface is straight alongshore and, depending on sand size, may have a low-tide terrace (fine sand) and/or a steeper, reflective high-tide beach with beach cusps (medium sand). The trough is usually drained by weaker rips that cross the bar every 250–500 metres. The deep trough and the presence of less obvious rips make this a particularly hazardous swimming beach.

In Australia, the longshore bars more commonly occur as second outer bars and troughs, particularly along the longer northern New South Wales and southeast Queensland beaches. There are only eight beaches of this type occurring solely as an inner bar, all along high-energy sections of the South Australian and Western Australian coast.

Higher waves tend to break continuously along the bar, often as heavy plunging breakers – that is, waves in which the crest suddenly thrusts forward onto the wave trough. In the vicinity of the deeper rip channels, waves may not break, and the deeper clearer water marks the location of the channel. The broken waves cross and pour off the bar, into the deeper longshore trough, where they quickly reform and continue shoreward as a lower wave to break on, or surge up, the beachface. The water that flows into the trough moves shoreward and returns seaward via two mechanisms: first, the water piles up along the beachface as wave set-up. As the water sets-down it moves both seaward as a standing wave and alongshore in both directions as rip-feeder currents. Second, as two adjacent and converging feeder currents merge they turn and flow seaward across the trough and bar as a rip current. The current pulses seaward with each set-down, with flow accelerating where it crosses the bar.

Rhythmic bar and beach

The rhythmic bar and beach **(RBB)** type is the commonest of the higher-energy beach types that occur around the southern Australian coast. There are a total of 279 RBB beaches across all southern states (see Table 5.2). These energetic beaches require two prime ingredients: relatively fine–medium sand (0.3 millimetres) and exposure to waves averaging more than 1.5 metres. Sydney's seven most exposed beaches are all RRB-type beaches: North Palm Beach, Avalon, Dee Why–Long Reef, Curl Curl, south Bondi, north Maroubra and Wanda–Elouera–North Cronulla.

The RBB type is characterised by an outer bar, which is separated from the beach by a deep trough. However, unlike the LBT type, the bar varies in width and elevation alongshore, with the shallower portions protruding landward and the deeper section seaward, giving the bar a rhythmic appearance (see Figures 5.6 and 5.9). Waves break more heavily on the shallower, shoreward-protruding bar sections, and together the broken wave and white water flow shoreward as a wave bore. The wave bore then flows off the bar into the deeper tough, where it moves shoreward

Figure 5.7 (left) A high-energy, south-coast dissipative beach, Dog Fence Beach in South Australia, with long Southern Ocean swell breaking over the outer bar, reforming in the central trough, then breaking across the inner bar, and up to 10 lines of breakers. Photo: A.D. Short

Figure 5.8 The beach at Point Lookout, North Stradbroke Island in Queensland, provides an example of a well-developed outer bar, wide deep trough and straight-curving beach. Photo: A.D. Short

and longshore as a rip-feeder current in a rip-feeder channel. Part of the wave reforms in the trough and breaks again on the shore. The water from both the wave bore and the swash piles up in the rip-feeder channel and moves sideways toward the adjacent rip embayment, which may be tens to more than a hundred metres alongshore. The feeder currents are weakest where they diverge behind the centre of the bar, but pick up in speed and intensity toward the rip embayment. In the rip embayment, the backwash returning down the beachface combines with flow from the converging rip-feeder currents. This water builds up close to shore (wave set-up), then pulses seaward as a strong, narrow (10–20 metres wide) rip current. The currents pulse every 30–90 seconds, depending on wave conditions. The rip current accelerates with each pulse and persists with lower velocities between pulses. Rip velocities are usually less than 1 metre per

second (3.5 kilometres per hour), but will increase up to 2 metres per second under higher waves.

The alternation between bar and rip in turn generates a longshore variation at the shoreline. Waves break more heavily on the shallower protruding bar sections, and white water pushes sand shoreward, extending the bar. In the deeper rip sections, waves may reach the shore unbroken, eventually to break as a strong plunging shorebreak, eroding the beachface. At the same time, the waves refract around the shallower sections to approach their lee at an angle. This combination results in a reworking of the shoreline, such that in lee of the rips the higher shorebreak erodes the beach and usually forms curved scarps, while in lee of the bar the lower-converging waves build out the beach. The alternating protruding and eroding sections are called megacusp horns and embayments, respectively, with the rip currents always originating in the embayment. The surf zone may be up to 100–150 metres wide, and the bar, rips and megacusps spaced every 250–500 metres alongshore.

To identify this beach type, look along the beach and you will see the distinctive horns and embayments, with a trough separating them from the bar sections off the horns, and deeper rip channels off the bay.

Transverse bar and rip

The transverse bar and rip (TBR) type is the most common and widespread of Australia's wave-dominated beach types. It occurs on 1028 beaches and occupies 16 per cent of Australia's sandy coast, predominantly around the southern coast (see Table 5.2). Many popular surfing beaches are of this type.

TBR beaches occur primarily on beaches composed of fine to medium sand (0.3 millimetres) and exposed to waves averaging 1.5 metres. This beach type received its name from the bars appearing transverse, or perpendicular, to and attached to the beach, separated by deeper rip channels and currents (see Figures 5.6 and 5.10). The bars and rips are usually regularly spaced, with a mean spacing ranging from as short as 150 metres on the lower-energy Northern Territory and Cape York Peninsula coasts, to 250 metres along the higher-energy New South Wales coast and 350 metres along the exposed south coast of Western Australia, with individual large rips spaced up to 500 metres and more apart. Their surf zones similarly range from 50 to 150 metres in width. The spacing and surf zone width increases with both increasing wave period and wave height.

The alternation of shallow bars and deeper rip channels causes a longshore variation in the way

waves break across the surf zone. On the shallower bars, waves break heavily, losing much of their energy. In the deeper rip channels they will break less and possibly not at all, leaving more energy to be expended as a shorebreak at the beachface. Consequently, across the inner surf zone and at the beachface there is an alternation of lower-energy swash in lee of the bars and higher-energy swash/shorebreak in lee of the rips.

This longshore variation in wave breaking and swash continues the beach reworking – which commences in the rhythmic bar and beach type – with slight erosion in the embayments and slight deposition in lee of the bars. This results in a rhythmic shoreline: building a few metres seaward behind the attached bars as deposition occurs, forming the megacusp horns and then being scoured

Figure 5.10 *Well-developed transverse bar and rip channels along Lighthouse Beach, New South Wales. Note the waves breaking on the bars, with no waves breaking in the deeper, darker rip channels. The rhythmic shoreline protrudes in lee of the bars and forms an embayment in lee of the rips.* Photo: A.D. Short

out and often scarped in lee of the rips forming the embayments.

The alternation of bars and rips also causes cellular circulation in the surf zone. Here, waves tend to break more on the bars and move shoreward as wave bores. This water flows both directly into the adjacent rip channel and closer to the beach, into the rip-feeder channels located at the base of the beach. The water in the rip-feeder and rip channel then returns seaward in two stages. First, water collects in the rip-feeder channels and the inner part of the rip channel, building up a hydraulic head (set-up) against the lower beachface. Once high enough, it pulses seaward (set-down) as a relatively narrow, accelerated flow – the rip current. The water usually moves through the rip channel, out through the breakers and seaward, as an expanding and decelerating current, for a distance up to twice the width of the surf zone. Some of this water may then be brought back onshore by the breaking waves.

The velocity of the rip currents varies with conditions. On a typical beach with waves less than 1.5 metres, they peak at about 1 metre per second, about walking pace. However, under high waves they may double that speed. What this means is that under average conditions, a rip may carry someone out from the shore to beyond the breakers in 20 to 30 seconds. Even an Olympic swimmer going at 2 metres per second would only be able to maintain his or her position, at best, when swimming against a strong rip. For these reasons, rip currents pose the greatest risk to beachgoers and are the cause of most beach rescues on the Australian coast.

Two other problems associated with rips and rip channels are their greater depth and their rippled seabed. The channel is usually 0.5 to 1 metre deeper than the adjacent bar, reaching a maximum depth of 3 metres. The faster seaward-flowing water forms megaripples on the floor of the rip channel. These are sand ripples 1 to 2 metres in length and 0.1 to 0.3 metre high that slowly migrate seaward. The combined effect is to produce a fast, seaward-flowing current in a deeper channel and with a softer undulating seabed – in stark contrast to the adjacent shallower, firmer bar where the waves flow shoreward. As a result, in the rip channel it is more difficult to maintain your footing for three reasons: the water is deeper, the current is stronger and the channel floor is less compact. Also, someone standing on a megaripple crest that moves or walks into the deeper trough may think the bottom has 'collapsed'. This may be one source of the 'collapsing sand bar' myth, an event that cannot and does not occur.

Low tide terrace

Low tide terrace (LTT) beaches are the lowest-energy intermediate beach type. They occur all around the Australian coast, although are more prevalent around the southern half of the continent. There is a total of 900 beaches of the LTT type (see Table 5.2). They tend to occur when waves average about 1 metre and sand is fine to medium, averaging 0.3 millimetres in diameter.

LTT beaches are characterised by a moderately steep beachface, which is joined at the low-tide level to an attached bar or terrace, hence the name. The bar usually extends between 20–50 metres seaward and continues alongshore, attached to the beach (see Figure 5.6 and 5.11). It may be flat and featureless, have a slight central crest (called a 'ridge') and may be cut every 100 to 200 metres by small, shallow rip channels, called mini-rips.

At high tide, when waves are less than 1 metre, they may pass right over the bar and not break until the beachface, which behaves much like a reflective beach. At spring low tide, however, the entire bar is usually exposed as a ridge or terrace running parallel to the beach. At this time, waves break by plunging heavily on the outer edge of the bar. Under typical mid-tide conditions, with waves less than 1 metre high, they break across the bar and a low surf zone is produced. The water is returned seaward, both by reflection off the beachface, especially at high tide, and via the mini-rips. If no rip channels are present, the rips flow across the bar as ephemeral, shallow transient rips.

Reflective

Reflective (R) sandy beaches lie at the lower-energy end of the wave-dominated beach spectrum. They are characterised by relatively steep, narrow beaches that are usually composed of coarser sand (mean= 0.4 millimetre). On the open Australian coast, sandy beaches require waves to be less than 0.5 metre in order to be reflective. For this reason they are also found inside the entrances to bays, at the lower-energy end of some ocean beaches and in lee of the reefs and islets that front many beaches. Reflective beaches are Australia's most common beach type, characterising 2670 beaches (25 per cent). They occur in every state, although they are more common around the southern half of the continent (see Table 5.2).

Reflective beach morphology consists of a narrow, steeper beach and swash zone, with beach cusps commonly present in the upper high-tide swash zone (see Figure 5.6). They have no bar or surf zone, as waves move unbroken to the shore, where they collapse or surge up the beachface (see

Figure 5.11 *A gradation from a low tide terrace cut by mini-rips (foreground) to a reflective beach (top) with no bar, at Number One Beach, New South Wales.* Photo: A.D. Short

Figure 5.12 *A steep, reflective beach with well-developed, high-tide beach cusps at Hammer Head, in the south of Western Australia.* Photo: A.D. Short

Figure 5.12). A surging breaker is one that breaks on a steep gradient over a very short distance (metres) and in doing so suddenly transforms into a wall of broken water that runs up the beachface.

The absence of surf and the steep beach is a result of four factors:

1. Low waves do not break until they reach relatively shallow water (< 1 metre).
2. The coarser sand results in a steeper gradient beach (5 to 10 degrees) and relatively deep nearshore zone (> 1 metre).
3. Because of the low waves and deep water, the waves do not break until they reach the base of the beachface.
4. Since the waves break at the base of the beachface, as collapsing or surging breakers, they must expend all their remaining energy over a very short distance.

Much of the energy goes into the wave run-up and backwash, while the rest is returned to sea as a reflected wave, hence the name.

The strong swash, in conjunction with the usually coarse sediment, builds a high, steep beachface, with beach cusps usually forming on the upper part of the beachface. Cusps are regular undulations in the beach consisting of a raised cusp horn and deeper cusp swale. The cusps are a product of a wave motion called sub-harmonic edgewaves. 'Subharmonic' means the waves have a period twice that of the incoming wave, while an edge wave is wave energy trapped against the beach that undulates along the shore and generates a cellular swash pattern that initiates the cusps. The edgewave period and the beach slope determine the length of the edgewave, which in turn determines the spacing of the cusps. Around southern Australia, where longer wave periods occur, cusp spacing can range from 20 to 40 metres, while in northern Australia, which has shorter waves, cusps are also shorter (10 to 20 metres).

Another characteristic feature of most reflective beaches is that all those containing a mix of sand sizes have what is called a beach step. The step is always located at the base of the beachface, around the low-water mark. It consists of a continuous band containing the coarsest material available, including rocks, cobbles, even boulders, and often numerous shells. The coarsest material is trapped at the base of the beachface by both the incoming unbroken wave and the high-energy swash and backwash. Because it is so coarse, its slope is very steep, hence the step-like shape. A beach step is usually a few decimetres in height, reaching a maximum of perhaps a metre. Immediately seaward of the step, the sediments usually fine abruptly and consequently assume a lower slope.

The combination of the steep beachface – a potential step-like drop-off into deep water – and the expending of all wave energy in a shorebreak (a wave that breaks suddenly at the shore as a plunging or surging wave) can produce hazardous conditions on reflective beaches, particularly when waves exceed 1 metre. The steep beach makes footing difficult, while the strong shorebreak can knock people over then drag them down the beach, and if carried past the step can deposit them in deep water.

Rip currents

Rip currents are a feature of most wave-dominated beaches and some tide-modified beaches. In this section we provide some further insight into the types, nature and occurrence of rip currents that are so prevalent – and pose such a hazard to beachgoers – around the Australian coast. To begin with, there are four types of rip currents: two associated with sandy beaches (called accretionary and erosional (beach) rips); one in which adjacent headlands, reefs or other structures play a major role (topographic rips); and one controlled by high waves and topography (megarips).

Feeder currents

A beach rip is a relatively narrow current of water flowing seaward through the surf zone. It represents a mechanism by which water brought onshore by breaking waves is returned to sea. Beach rips originate close to shore as the shoreward-moving, broken waves and wave bores are deflected alongshore at the shoreline. This water moves along the base of the beach as a rip-feeder current. The two feeder currents arriving from opposite directions usually converge in the rip embayment, turn and flow seaward across the bar and through the surf zone as the rip current. The feeder currents may maintain a deeper rip-feeder trough close to shore, and a deeper rip channel may be cut into the bar and through the surf zone.

The converging feeder currents turn, accelerate and flow seaward through the surf zone, as a relatively narrow 10–20-metre-wide flow, either directly or at an angle, at speeds up to 2 metres per second. As the confined rip current exits the surf zone and flows seaward of the outer breakers, it expands and may meander as a larger rip head (see Figure 5.13). The rip current's speed decreases and will usually dissipate within a distance of one to three times the width of the surf zone.

Figure 5.13 *A rip current flowing beyond the surf zone as a rip head, with an earlier rip head meandering to right, Miles Beach, Bruny Island, Tasmania.* Photo: A.D. Short

Beach rip currents exist in some form on all beaches where there is a surf zone, particularly when waves exceed 1 metre. Their formation depends on wave breaking, while their spacing is more a function of wave period and height.

Accretionary rips

Accretionary rips are associated with beach recovery, onshore sediment transport and waves usually less than 1.5 metres on a swell coast. They usually have velocities of 0.5–1 metre per second, but can attain velocities of 1.5 metres per second (that is, 5.4 kilometres per hour), particularly as the flow becomes more confined to a narrow rip channel and towards low tide.

Accretionary beach rips are the most common rip type encountered on Australian beaches. They pose the greatest risk to beach users, because waves are relatively low and conditions apparently normal, often with 'calm', non-breaking water in the rip channel (see Figures 5.10 and 5.11). They range in spacing from 100–150 metres apart in the short seas of northern Australia, to 250 metres on the southeast coast, to as much as 350 metres along the exposed southern coast, with maximum spacing reaching 500 metres and more. On an average day there are 13 500 accretionary beach rips operating around the Australian coast. The rips and adjacent bars tend to be relatively stationary because the rips occur during periods of decreasing to lower wave conditions. In a transverse bar and rip beach type, the bars and rips can remain relatively fixed in location under suitable waves for days to weeks, as the bars migrate shoreward and the rip channels gradually infill. They will, however, skew and migrate longshore at up to a few metres per day when exposed to persistent oblique waves.

Erosional rips

Erosional rips are associated with increasing wave height (>1.5 metres) and beach erosion, with waves usually breaking across the rip current. They are hydraulically controlled by the prevailing wave and beach conditions and, as such, will change in location, spacing and strength as conditions change. Consequently, erosional rips can be highly mobile and transient in nature, and are one cause of the so-called 'flash' rips. As wave height increases the erosional rips become larger, more widely spaced, fewer and stronger, commonly reaching several hundred metres in spacing. During this process they are a major cause of beach erosion and conduit for offshore sediment transport, with the sediment deposited by the expanding rip heads beyond the breaker zone (see Figure 5.14). Because erosional rips accompany rising seas, they tend to discourage bathers from entering the surf and consequently provide a self-mitigating process to reduce their

Figure 5.14 *Three regularly spaced, erosional beach rips (arrows) with large rip heads along Surfers Beach, in the south of Western Australia.* Photo: A.D. Short

hazard. Little field data is available on erosional rips, but their velocities generally exceed those of accretionary rips and are expected to be in the range of 1.5–2 metres per second (5 to 7 kilometres per hour). Erosional rips occur on all beaches that have accretionary rips, whenever higher wave conditions occur. However, because of their wider spacing they are larger but fewer in number, compared to the accretionary rips, and because they are more transient they are harder to spot.

Topographic rips

A topographic rip is a rip current whose flow is partly controlled by a topographic feature, usually a solid structure such as a headland, reef, groyne or jetty, and in some cases a prominent sand feature. Topographic rips occur wherever such a feature is present in a surf zone. The rips are initiated by normal wave breaking and onshore water flow. However, as the surf zone and feeder currents move longshore they are deflected seaward along the side of the topographic feature. As a consequence, topographic rips are fixed in location, always flowing out along the side of the same structure. This is why some have been given local names, such as The Alley at North Narrabeen and the Backpacker Express at South Bondi.

The velocity of topographic rips depends on wave conditions, and they can range from a weak flow in a shallow, narrow channel during low waves, to a deep (>3-metre) channel with wider, strong flows (5–10 kilometres per hour) during high waves. Compared to beach rips, topographic rips usually have a stronger flow, a deeper channel and flow further seaward (see Figure 5.15). They also have the additional hazards of the adjacent structure, such as rocks and reefs.

Topographic rips are very common, particularly around the higher-energy southern Australian coast. The predominance of relatively short, headland-bound beaches results in approximately 4000 topographic rips operating on an average day, most occurring against headlands and reefs on embayed beaches. They are therefore a major hazard on our beaches. It is best to avoid them by not swimming near headland rocks and reefs. Experienced surfers, however, find they provide an excellent access to the outer surf zone.

Megarips

Megarips are large-scale topographic rips that occur in association with high wave conditions. They tend to form when waves exceed ~3 metres on embayed beaches up to 3 to 4 kilometres in

Figure 5.15 *A well-developed topographic rip and rip head at Point James, South Australia.* Photo: A.D. Short

length. As wave conditions increase, the erosional rips increase in spacing and eventually merge with adjacent topographic rip or rips, to form one or two large-scale megarips and draining an entire beach system (see Figure 5.16). In some cases, megarips can drain a series of smaller embayments linked by a common surf zone and driven by one large rip, as occurs at Sydney's Mckenzies, Tamarama and Bronte beaches.

Megarips reach velocities of up to 3 metres per second (10 kilometres per hour) and can flow more than 1 kilometre longshore and up to 1 to 2 kilometres seaward. As a consequence, they

Figure 5.16 *A megarip (arrow) flowing out of the centre of Dee Why Beach, New South Wales.* Photo: A.D. Short

can transport more sand and coarser sand further seaward, which results in faster and more severe beach erosion, particularly in lee of the rip location. Because the sand is transported further offshore and to greater depths, the result is slower beach recovery compared to straight beaches, particularly following severe storms. In extreme conditions, sand is lost permanently to the inner shelf, which off Sydney has led to the accumulation of the massive inner-shelf sand deposits 1 to 3 kilometres offshore.

Rip occurrence

Beach rips operate on all intermediate beaches during normal wave conditions and are the most commonly encountered rip type. Topographic rips also occur under normal wave conditions but intensify during rising seas. Erosion rips operate only during rising and high seas, and are associated with general beach lowering and erosion. Megarips occur only on embayed beaches or beaches influenced by permanent structures during periods of high waves (> 3 metres).

Rip currents are therefore an integral part of wave-dominated, intermediate beaches and some tide-modified beaches. Australia has 2246 wave-dominated intermediate beaches, all with rip-dominated cellular circulation, together with 933 tide-modified beaches, which usually experience rips at low tide. These 3179 beaches represent 30 per cent of Australia's 10 685 mainland beaches. The remainder tend to be free of rip circulation, particularly the tide-dominated beaches, which make up 34 per cent of the beaches. The rip-dominated beaches occur particularly around the wave-dominated southern Australian coast. The tide-modified beaches with rips tend to occur along the more exposed sections of the Northern Territory, western Gulf of Carpentaria and eastern Queensland coast.

Rip spacing

The size and spacing between rip currents depends on wave conditions. They can be located as close as every 50 metres, to large rips more than 500 metres apart. The shortest rips occur in northern Australia, where short seas with 3 to 5-second wave periods result in more closely spaced rips. Mean rip spacing ranges from 100 metres in the Kimberley region, to 110 to 140 metres in the Northern Territory, to 150 metres on Queensland's Cape York Peninsula and 170 metres along the eastern Queensland coast, in lee of the Great Barrier Reef. These contrast with the southern Australian beaches exposed to longer period (10 to 14-second) and higher swell conditions, where mean rip spacing ranges from 250 metres in New South Wales to 350 metres in southern Western Australia, with individual high-energy beaches having rip spacing in excess of 500 metres. As a consequence, northern Australian, tide-modified beaches have 4650 rips, which occupy 750 kilometres of the coast with an average spacing of 160 metres and rip density of 6.25 rips per kilometre, while southern Australia's wave-dominated beaches have 8950 beach rips occupying 2480 kilometres with an average spacing of 280 meters and density of 3.6 rips per kilometre.

In addition to the above accretionary beach rips, there are the 3965 topographic rips around the Australian coast. Most (3630) occur around the southern half of the continent, with an additional 335 occurring around the northern half. Their lesser occurrence in the north is due to the low waves, rather than the lack of topographic features. The prevalence of topographic rips around the Australian coast is due to two factors. First, the relatively short average length of the beaches (mean = 1.37 kilometres) suggests that most beaches are embayed and bounded by topographic features such as headlands and reefs; and second, the prevalence of regular high waves (>1.5 metres) around the southern coast, which is required to generate the surf zone and rips. While northern Australia has equally short beaches and topographic boundaries, it lacks the high waves and surf needed to drive such rips under normal wave conditions. They do, however, occur during higher cyclonic waves.

When the two rip types are combined on a normal day there are approximately 13 600 accretionary beach rips and 4000 topographic rips operating around the Australian coast, a total of 17 600 rip currents (see Table 5.3). During higher waves this number will decrease as erosional rips dominate with fewer large rips, and as megarips replace both beach and topographic rips. For example, along Sydney's 36 ocean beaches, there are usually 86 beach rips and 36 topographic rips operating under normal conditions, with waves averaging 1.5 metres. When waves exceed 3 metres these 122 rips increase in spacing, merge and are replaced by just 24 large megarips.

Beach change

The above changes in rip spacing suggest that when wave conditions change, beaches respond by eroding during higher waves and recovering during periods of lower waves, with sand moving onshore

State/region	Beach rips	Mean spacing (m)	Standard deviation (σ) spacing (m)	Topographic rips
Cape York	1598	145	77	10
East Queensland	2544	173	42	105
New South Wales	2952	246	63	677
Victoria	2370	257	95	758
Tasmania	1112	282	106	501
South Australia	1008	305	122	791
Western Australia	1470	355	117	903
Kimberley	15	100	–	3
Northern Territory	499	139	35	217
Total	13568	100–355		3965

Table 5.3 Australian rip current type, distribution and spacing during average wave conditions

and building the beach. Therefore, while each beach will have an average modal type, which represents its typical condition under average wave conditions, as illustrated in Figure 5.6, so too, all beaches can shift temporarily to higher or lower-energy beach types as waves change. This is why most wave-dominated beaches are in a continual state of flux, as they adjust to the ever-changing wave conditions.

Tide-modified beaches

Whereas the southern half of the continent is dominated by high waves and low tides that result in predominantly wave-dominated beaches, the opposite occurs across the northern half of the continent. Waves at the shore are generally low across the top, averaging 0.1 metre in the Kimberley, 0.3 metre in the Northern Territory and 0.4 metre on Cape York Peninsula, while tides are greater than 2 metres right across the Top End, reaching several metres in central Queensland, Torres Strait and across the entire northwest Kimberley region. As a consequence, 56 per cent of beaches in northern Australia are tide-dominated, 19 per cent are tide-modified, only 7 per cent are wave-dominated and 18 per cent fronted by rocks or reefs. By contrast, in southern Australia a total of only 14 per cent of beaches are tide-modified or tide-dominated.

By definition, tide-modified beaches occur when the tide range is between three and 10 times the wave height. Where tides remain low (1 metre or less), such as in Shark Bay, the waves must be less than 0.3 metre to be considered tide-modified, while across most of the Top End the increasing tidal range, as well as lower wind waves, produces these

7. Reflective + low tide terrace (+rips)

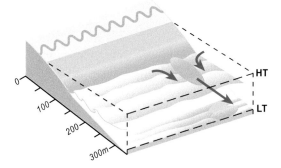

8. Reflective + low tide bars & rips

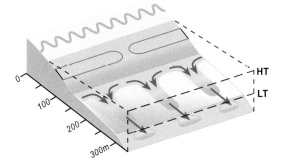

9. Ultradissipative

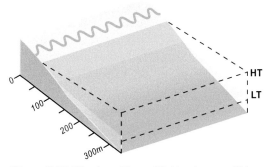

Figure 5.17 The three tide-modified beach types. All have high-tide beach cusps (curving orange line), possibly rips at low tide (blue arrows) and a wide intertidal zone between high tide (HT) and low tide (LT).

conditions on the open coast. Tide-modified beaches consist of three beach types, with increasing wave energy: reflective plus low-tide terrace, reflective plus bar and rips, and ultradissipative (see Figure 5.17).

Reflective plus low tide terrace beaches

The reflective plus low tide terrace (R+LTT) beach type is the most common tide-modified beach occurring on 754 beaches, predominantly on longer Cape York Peninsula–east Queensland beaches, where they average 4 kilometres in length; in the Northern Territory, where the beaches average 1.6 kilometres in length; and a few shorter Kimberley beaches. They are also the most common tide-modified beach to occur along sheltered higher-tide range sections of the Victorian, Tasmanian and South Australian coasts, with 100 located in these locations (see Table 5.2).

R+LTT beaches are characterised by a relatively steep, cusped reflective high-tide beach, usually composed of medium sand (0.45 millimetres) exposed to short-period waves averaging 0.45 metres in height, with tidal range averaging up to 10 times (4.5 metres) the wave height. The beachface slopes to low tide, where it abruptly grades into a low-gradient, low-tide terrace, usually composed of finer sand, which can extend tens of metres seaward (see Figures 5.17 and 5.18). At high tide, waves pass unbroken over the terrace and only break on reaching the high-tide beach, similar to the reflective wave-dominated beach. As the tide falls, waves increasingly begin to break across the terrace, and at low tide break on the outer edge to produce a wide, shallow surf zone across the terrace. If rips are present, they are only active at low tide and may cut a channel across the terrace.

Figure 5.18 A steep, reflective high-tide beachface fronted by a 100-metre-wide, low-tide terrace and crossed by shallow drainage channels, North Harbour Beach, Mackay, Queensland. Photo: A.D. Short

Reflective plus low tide rips

The reflective plus low tide rips (R+LTR) beach type occurs on 179 beaches, primarily on Cape York Peninsula–east Queensland and in the Northern Territory, with only three occurring in the Kimberley and three in southern Australia. These are the highest-energy of the tide-affected beaches, with waves averaging 0.7 metre, medium sand (0.35 millimetre) and tides averaging 2.5 metres (RTR=3).

The main environmental difference between R+LTR and ultradissipative beaches is the coarser sand. At high tide, the waves pass over the bar without breaking until they reach the beachface, where they usually maintain a relatively steep beach with cusps. As the tide falls, waves begin breaking on the bar. At low tide there is sufficient time and wave energy to generate the rips and scour rip channels (see Figures 5.17 and 5.19), which have an average spacing of 140 metres (σ=35 metres), the same spacing as the wave-dominated TBR in northern Australia. The R+LTR tend to occur on longer beaches, with average lengths in the Kimberley, Northern Territory and Cape York Peninsula of 2.4 kilometres, 1.5 kilometres and 5.6 kilometres, respectively. These and the TBR provide the best surfing beaches in northern Australia, although the rips do pose a hazard to swimmers. Consequently, together with TBR beaches, they are also physically the most hazardous beaches in northern Australia, particularly at low tide, when the rips dominate.

Figure 5.19 *A low tide bar and rip beach at low tide on Northdown Beach, northern Tasmania. Note the narrow, straight high-tide beach, wide intertidal zone and breaking waves and rip channels in the low-tide zone.* Photo: A.D. Short

Ultradissipative beaches

Ultradissipative (UD) beaches occur in higher-energy, tide-modified locations, where waves average 0.6 metre and the beaches are composed of fine sand (mean=0.25 millimetre). UD beaches are characterised by a very wide (200 to 400 metres) intertidal zone, with a low to moderate gradient, high-tide beach and a very low gradient to almost horizontal low-tide beach. Because of the low gradient right across the beach, waves break across a relatively wide, shallow surf zone as a series of spilling breakers (see Figure 5.20). The wide, spilling surf zone dissipates the waves so effectively that they are

Figure 5.20 *A wide, ultradissipative beach at low tide in Stanage Bay, Queensland.* Photo: A.D. Short

10. Beach + ridged sand flats

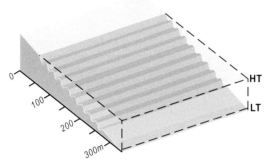

11. Beach + sand flats

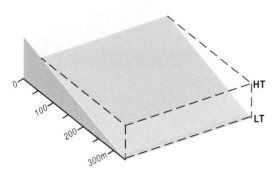

11 & 12. Beach + tidal sand/mud flats

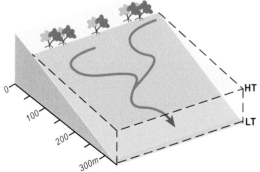

Figure 5.21 *The four tide-dominated beach types. All usually have a small high-tide beach, which abruptly grades into the wide sand flats. Mangroves may grow on tidal sand and mud flats (12 & 13).*

known as 'ultradissipative' beaches. During periods of higher waves (>1 metre), the surf zone can be well over 100 metres wide, although still relatively shallow. Basically, the fine sand induces the low gradient, while every 12 hours the tidal range moves the higher waves backwards and forwards across the wide intertidal zone. The two act to plane down the beach, and the lack of stability of the surf zone and shoreline prevent the formation of bars and rips.

There are 180 UD beaches, predominantly on eastern Cape York Peninsula–east Queensland (see Figure 5.20), the northwest of Western Australia, including Eighty Mile Beach and Broome's Cable Beach, several in the Northern Territory and eight along the higher-tide range along the northern Tasmanian coast (see Table 5.2).

Figure 5.22 *Ridged sand flats at Hardewicke Bay in Spencer Gulf, South Australia (8 ridges).* Photo A.D. Short

Tide-dominated beaches

Tide-dominated beaches are the most prevalent beach type across northern Australia, with a total of 2940 (56 per cent) spread across all three regions (see Table 5.2). The combination of higher tides, low waves, and in places abundant fine sand and mud, all contribute to their dominance. In the Kimberley, where they are most prevalent, the highly indented nature of the coast provides additional sheltering and lower waves, which combine with the high tidal ranges to produce this beach type.

Due to the irregular rocky shoreline, all the tide-dominated beaches in the Kimberley average only 400 to 500 metres in length. Each of these beaches has a reflective high-tide sand beach, with the intertidal zone grading from more exposed, ridged sand flats through to mud flats (see Figure 5.21). Most of the flats are flanked by tropical seagrass meadows below low water, while mangroves grow in the intertidal zone of many of the lower-energy systems. They also occur in sheltered gulfs and bays in southern Australia, with a total of 650 (12 per cent) occurring in the south, primarily in Shark Bay, the South Australian gulfs, Victoria's Port Phillip Bay and sheltered southern Tasmanian bays.

Beach plus ridged sand flats

The beach plus ridged sand flats (R+RSF) is the highest energy of the tide-dominated beaches. They tend to occur where waves average 0.5 metre and tides average 4.5 metres, resulting in an RTR of 9. They are characterised by a relatively steep, occasionally cusped high-tide beach, which abruptly grades into a very low gradient, sandy intertidal zone, covered by regularly spaced low amplitude (5 to 10 centimetres), shore-parallel sand ridges (see Figures 5.21 and

Figure 5.23 *A narrow high-tide beach fronted by 1-kilometre-wide sand flats in upper Spencer Gulf, South Australia.* Photo: A.D. Short

5.22). In northern Australia, the intertidal flats average 600 metres in width and have on average seven ridges with a mean spacing of 80 metres. The flats can range from 50 to 5000 metres in width, with up to 22 ridges, and at spring low tide they are fully exposed. Little is known about the morphodynamics of the ridges, although it is assumed they are formed by waves breaking and reforming over the ridges as the tide falls. There are a total of 390 beaches with ridged sand flats that are most prevalent around Cape York Peninsula and the Northern Territory, with 70 R+RSF beaches also occurring in the South Australian gulfs and southern Tasmanian bays.

Beach plus sand flats

The beach plus sand flats (R+SF) is the most common beach type in northern Australia, with a total of 1053 beaches (30 per cent). Although most occur in the

Figure 5.24 *Two-kilometre wide, tidal sand flats front the narrow high-tide beach at Port Smith, Western Australia.* Photo: A.D. Short

Figure 5.25 *A high-tide sand beach, fronted by 200-metre-wide, intertidal mud flats at Hut Point, in the eastern Northern Territory.* Photo: A.D. Short

Kimberley and Northern Territory, there are also 513 beaches of this type in southern Australia, primarily in the South Australian gulfs and bays (see Table 5.2). They are similar to the R+RSF, except that waves are lower (mean 0.26 metre), tides higher (mean 5 metres), resulting in an RTR of 20. R+SF beaches have a relatively small, steep high-tide beach, which grades abruptly into intertidal sand flats that average 300 metres in width (range 10–3000 metres) (see Figure 5.21). The sand flats are low and featureless (see Figure 5.23), apart from small wave ripples, indicating that wave energy is still sufficiently high to mobilise sediment across the flats, but not high enough to form the ridges of the previous beach type.

Beach plus tidal sand flats

The two lowest-energy beach types occur in areas with lower waves and where the tidal range is more than 30 times the wave height. The beach plus tidal sand flats (R+TSF) have a small, reflective, steep and usually coarse-grained high-tide beach, fronted by intertidal sand flats averaging 350 metres in width (range 50–2500 metres) (see Figures 5.21 and 5.24). They differ from the R+SF in that they receive lower waves (mean 0.16 metre) with similar tides (mean 5 metres), producing an average RTR of 32. Waves are sufficiently low, and tidal energy sufficiently high, for the tidal currents to imprint themselves on the tidal flats, and in some locations for mangroves to colonise the upper intertidal zone. These are the second-most common beach type in northern Australia, with a total of 764 (15 per cent) occurring in each region, but particularly in the Kimberley, together with 70 in South Australia (see Table 5.2). Many of these flats grade from inner sand flats to outer mud flats, with the sand averaging 300 metres wide and the mud extending out on average to 500 metres.

Beach plus tidal mud flats

The final tide-dominated beach type is the beach plus tidal mud flats (R+TMF). These occur in similar wave-tide regimes to the R+TSF, with waves also averaging 0.16 metre, but tides 8 metres or higher, providing an average RTR of around 50. However, in addition to the low waves there is usually a local source, such as a river, to supply the mud to the shoreline. R+TMF have a low, narrow high-tide beach, usually composed of coarse shelly sand (a chenier), which grades often very abruptly into wide, very low gradient intertidal mud flats, with mangroves often colonising the upper intertidal (see Figures 5.21 and 5.25). They average 500 metres in width, ranging from 50–2000 metres. There are 240 (5 per cent) of these beach types across northern Australia, with most occurring in the Kimberley and Northern Territory, while there are none in southern Australia.

Beaches with rock or reef flats

The 13 beach types discussed above cover 90 per cent of Australian beaches. They also represent beaches that, although often bounded by headland and rocks, usually have beach and surf zone tidal flats free of rocks and reefs. There are, however, two additional beach types that are substantially influenced by the presence of rocks and reef flats that extend seaward from the base of the beach.

14 & 15. Reflective + rock/reef flats

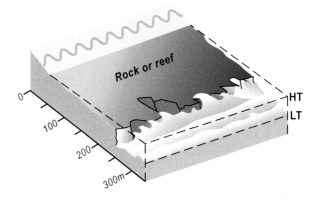

Figure 5.26 *A high-tide sand beach, fronted by a rock or reef flat.*

Beaches fronted by intertidal rock flats

Beaches fronted by intertidal rock flats (R+RF), and in some cases supratidal rock platforms, occur in the Kimberley (172), the Northern Territory (402), where most are fronted by laterite reef flats, and on Cape York Peninsula (57); in all, these comprise 18 per cent of the northern Australian beaches. There are 66 R+RF beaches in Victoria and some in the south of Western Australia – the southern beaches all fronted by rock flats.

This beach type is dependent more on local geology than wave-tide processes, and they can occur in both low and high wave and tide environments. The intertidal rock flats average 270 metres in width and range from 50 to 3000 metres wide (σ=330 metres). They usually consist of a steep, sandy high-tide beach, with the rocks extending seaward from the base of the beach (see Figure 5.26). The beaches tend to be relatively short, ranging between 400 and 800 metres in length, and are usually bounded by rock headlands or reefs. At high tide, waves break heavily across the rock flats, producing treacherous surf conditions and some of Australia's most hazardous beaches, particularly along the high-energy southern coast. At low tide, depending on the elevation of the flats, the waves may only break on the outer edge of the rocks, with a bare intertidal rock surface and no wave reaching the high-tide beach (see Figure 5.27).

Fringing coral reef beaches

A beach type that can only occur in northern Australia is the beach fronted by a fringing coral reef (R+CR). There are a total of 218 (6 per cent) of these beaches, primarily along Western Australia's Ningaloo (36) and Kimberley (163) coasts, together with 35 in the Northern Territory and 20 along eastern Cape York Peninsula. R+CR beaches are also characteristic of the numerous sand cays and reef islands that occur on offshore coral reefs (see Chapter 8). The beaches consist of a usually steep, high-tide reflective beach, often composed of coarse coral fragments, fronted by the reef. The reef flats average 300 metres in width but can range from 50 to 2000 metres. In the Kimberley, these are the shortest beach type, averaging only 300 metres in length (see Figure 5.28). At low tide, the reef is usually exposed and waves break on the outer reef edge, while at high tide the waves break across the reef flat and reach the beach.

The 15 beach types discussed above include every one of Australia's more than 10 000 mainland beaches, as well as those on many islands not counted in this total. Table 5.2 lists their distribution by state, and Figure 5.3 illustrates the general distribution around the entire coast. As would be expected, the

Figure 5.27 *Coal Point beach in Victoria consists of a narrow, sandy high-tide beach, fronted by 100-metre-wide intertidal rock flats.* Photo: A.D. Short

Figure 5.28 *Coral reef fringing the beach at Turquoise Bay, Western Australia.* Photo: A.D. Short

southern coast is dominated by wave-dominated beaches, the northern coast by tide-modified and tide-dominated beaches, with Western Australia, which straddles both zones, having an equal share of wave and tide-influenced beaches. The list of beach guides provided in Further Reading (see page 281) offers a comprehensive source for finding out the details of each and every Australian beach, its location, nature and type.

Overview of the beaches of Australia

In this section we take a brief tour around the Australian coast, providing an overview of the beaches of each state and the Northern Territory,

Figure 5.29 One of Queensland's more interesting beaches is Frangipani Beach at Cape York, with the cape protruding to the right of the ridged sand flats of the beach. Photo: A.D. Short

including photographs of some of the more interesting beaches.

Queensland

Queensland has 1650 predominantly low to moderate-energy, tide-dominated and tide-modified beaches spread along three coasts: the Gulf coast, the reef-protected east coast and the exposed southeast. The low-energy Gulf coast has Australia's most extensive tide-dominated mud flats in the south and generally long sand/shelly tide-modified beaches along the western side of Cape York Peninsula. All of these beaches are interrupted by numerous rivers and creeks, and a few rocky sections are composed of bright-red laterite bluffs. Cape York Peninsula is capped by the famous Frangipani Beach (see Figure 5.29), the northernmost in Australia.

Much of the east coast is located in the lee of the Great Barrier Reef, which is why there are relatively low waves at the shore. The high rainfall maintains numerous rivers and streams that deliver an abundance of sediment to the coast, while numerous headlands, capes and islands occupy the shore. As a consequence, the beaches range from sheltered, tide-dominated flats in lee of the capes, to more exposed, wave-dominated beaches facing into the southeast trade winds and waves. The latter are also backed in places by some of the largest tropical sand dune systems in the world, all usually covered in dense tropical vegetation.

The southeast coast of Queensland is free of the reef and exposed to the Southern Ocean swell. It is a moderate to occasionally high-energy coast, containing most of Queensland's 94 energetic wave and rip-dominated beach systems. The coast also lies at the terminus of a littoral cell that has been delivering quartz sand from northern New South Wales rivers via the beaches to supply the world's largest series of sand islands – South Stradbroke, North Stradbroke, Moreton, Bribie, Cooloola and the massive Fraser Island, which hosts Queensland's longest beach (see Figure 1.1). Likewise, the famous Gold Coast–Surfers Paradise region has a series of exposed rip-dominated beaches, a paradise for surfers, but hazardous to other beachgoers.

New South Wales

New South Wales beaches are controlled by their geology. They average 1.35 kilometres in length,

with the longest being only 29 kilometres long, and most are bounded by headlands. The 755 open coast beaches are exposed to a persistent, moderate southeast swell and tides less than 2 metres, which maintain exclusively wave-dominated and predominantly rip-dominated beaches (see Figure 5.30). On average, there are 2952 beach rips operating along the coast, the greatest number of any state. The exposed beaches are spread along the entire coast; they tend to be longer in the north, and in a few places are backed by extensive dunes. Because of the dominance of headlands and the dominant southerly swell, the southern end of many beaches is partly sheltered by a southern headland, resulting in lower-energy southern sections (see Figure 5.11) containing reflective to low-tide terrace

Figure 5.30 In northern New South Wales, Hat Head Beach, with the head in the background, is a typical rip-dominated beach, with wave energy decreasing in the lee of the head. Photo: A.D. Short

Figure 5.31 West Wingan Beach, in Croajingolong National Park, Victoria, is partially backed by high, vegetated sand dunes. Explorers Bass and Flinders sought shelter in the southern corner, known as Fly Cove, in 1798. Photo: A.D. Short

beaches. As wave energy increases along many of the beaches, the beach grades into a rip-dominated, central–northern section. The rip-dominated Sydney beaches are typical of the coast. Forty-three per cent of New South Wales beaches are located in national parks and reserves, and thus are preserved in their natural state.

Victoria

Victoria has four coasts: the east, central, Port Phillip and the wild west. At Cape Howe, on the border with New South Wales, the geology, coast and beaches change dramatically. The eastern Victoria coast (Cape Howe to South Point) faces into Bass Strait and its strong gales, with the Gippsland Basin fronted by a long, open coast with few rocky sections. These combine to produce the long Ninety Mile Beach and the equally extensive lake systems that back it. These are bordered to the north by the more rock-dominated beaches of Croajingolong National Park (see Figure 5.31) and in the south by the rugged granite of Wilson's Promontory National Park. The central coast (South Point to Cape Otway) also faces Bass Strait and receives swell from the Southern Ocean, resulting in a series of curving, wave-dominated beaches interspersed with shorter, rock-bound beaches, especially between Torquay and Cape Otway. Port Phillip has 232 lower-energy beaches, which are more exposed on the eastern shore while very sheltered on the lower western shore. The west Victorian coast faces the winds and waves of the Southern Ocean. It begins with the limestone-dominated Port Campbell coast, then expands into a series of long bays west of Warrnambool that have curving, exposed beaches, and culminates in the long, high-energy, rip-dominated Discovery Bay and its

backing massive dunes.

Tasmania

As well as several large islands, Tasmania has four distinctive coasts:

- the northern Bass Strait coast, with its higher tides and windy seas
- the moderate-energy east coast
- the bays and capes of the south coast
- the wild wind and wave-battered west coast.

The north coast ranges from sheltered, tide-dominated and modified beaches in the west to more exposed, rip-dominated beaches to the east (see Figure 5.32), the latter backed by dunes blowing inland for kilometres. The east coast resembles the New South Wales south coast, with numerous embayed wave and rip-dominated beaches and a series of stunning national parks, while the south coast between Cape Pillar and South East Cape ranges from exposed headlands with occasional beaches to predominantly sheltered beaches and sand flats of the bays. The west coast is exposed to the most energetic wave climate in the world, with waves averaging 3 metres, but reaching up to 18 metres. The coast is a mix of battered rocks and remote, high-energy, rip-dominated beaches.

South Australia

South Australia has a long and variable coastline. While the south and west are exposed to high-energy wave and winds, the more sheltered central gulfs and northern Kangaroo Island produce a series of tide-modified and some tide-dominated beaches.

The 400-kilometre-long southeast coast of South Australia (see Figure 5.33) is wave-dominated and includes Australia's longest continuous beach – the Coorong, which is also one of the highest-energy beaches in the world. The central Fleurieu and Yorke peninsulas, Kangaroo Island and St Vincent and Spencer gulfs provide a wide range of beaches, from exposed high-energy to very sheltered, tide-dominated beaches in the upper gulfs. The western Eyre Peninsula, beginning at the aptly named Cape Catastrophe, is an arid, highly exposed and predominantly high-energy coast. It features massive rip-dominated beaches that are backed by equally massive dune systems and interspersed with several sheltered bays.

Western Australia

Western Australia has 10 194 kilometres of coast and 3426 beaches, which equates to one-third of Australia's coast and beaches. The coast spans 17 degrees of latitude, from the tropics to the temperate south. As a consequence, the beaches range from the cooler, wave and rip-dominated south, to the tropical, tide-dominated north. Roughly half the beaches are wave-dominated and half are tide-dominated. In addition, there are smaller percentages of tide-modified beaches and Australia's largest collection of beaches fronted by fringing reefs and rock flats, predominantly in the northwest. The south coast beaches contain most of Australia's highest-energy, and most hazardous, beaches, usually backed by massive dune systems (see Figure 5.34). The heel between Cape Leeuwin and Cape Naturaliste is a high-energy, calcarenite-capped granite coast, which contains some of Australia's best surf and is

Figure 5.32 *Ringarooma Bay, Tasmania, is exposed to westerly waves and winds, which result in an energetic, double-bar surf zone backed by massive coastal dunes.* Photo: A.D. Short

Figure 5.33 *The curved and sheltered Nora Creina bay and beach contains a protected beach and one of the few settlements in South Australia's southeast. It is located within a breach, in a partly submerged Pleistocene (calcarenite) barrier.* Photo: A.D. Short

139

Figure 5.34 *On the south coast, long Southern Ocean swell rolls across the pure white sands of Wharton Beach, in La Grande National Park, Western Australia. A beautiful beach, but beware of strong rips.* Photo: A.D. Short

predominantly national park. The west coast up to North West Cape is to a large degree sheltered behind submerged Pleistocene calcarenite barriers (reefs and islands), and further north by fringing coral reefs, resulting in lower waves and quieter beaches, while the strong southerly winds have built substantial dunes in places. Lithified calcarenite dunes dominate the 360-kilometre-long Zuytdorp Cliffs, which partly protect the sheltered, hypersaline Shark Bay and its very low-energy calcareous beaches and extensive seagrass meadows. The Pilbara coast is dominated by

gradually increasing tides and wide beaches backed by extensive salt flats. While the rugged Kimberley coast has more rock than beaches, the beaches are restricted to generally short, sheltered and tide-dominated systems, many fringed by coral reefs.

Northern Territory

The Northern Territory has a 4500-kilometre-long coast and 1488 beaches, all located in the tropics and all potentially home to sunbaking crocodiles. Because of the generally low waves and high tide, these beaches are predominantly tide-dominated (53 per cent), with a substantial number fronted by rock flats, usually laterite, and a few wave-dominated on the more exposed eastern Arnhem Land coast. There is considerable variability around the coast, with very low-energy beaches with sand and mud flats and extensive mangroves in the west, and a mix of beach systems, rocky coast, bays and islands along the north and Arnhem Land coast (see Figure 5.35) and south into the Gulf as far as Groote Eylandt. The southern Gulf coast is a very low-energy tide-dominated and mangrove-fringed shoreline.

Conclusion

Australia is renowned for its beaches. However, can this claim be justified, given that most Australians and visitors see or know of only a tiny fraction of our more than 10 000 beaches? The short answer is a definite 'yes'.

What makes Australian beaches so good? The answer is a combination of six factors related to geology, climate, oceanography and marine biota.

Geology provides two factors: the number of headlands that often form attractive boundaries, backdrops and viewing platforms to many of the beaches, such as at Noosa, Bondi, Bells, Waitpinga, Cottesloe and Mindel beaches, and the fact that it helps us to divide the coast into the more than 10 000 beaches. The usually clean, white-to-yellow quartz sand that stems from ancient granite rocks, which is the source of half of Australia's beach sand, is another.

Around half the coast and on many coral sand cays the marine biota provides predominantly white carbonate detritus to form equally clean white beaches. The carbonate material is derived in the south from the inner continental shelf, in quieter environments from seagrass meadows and in the north from fringing coral reefs. Due to its predominantly arid climate, half the continent has no rivers and has had none for the past 40 million years. This means that little fine silt and clay are delivered to the southern coast to discolour the water. The result is clean, clear water over the clean, white-yellow sands.

Climate, via oceanography, provides the waves to roll into the beaches, with the southern coast dominated by regular surf year-round. Finally, climate, through wind, is also responsible for the extensive dunes that back most of the beaches, including the longest and largest coastal dune systems in the world.

The southern Australian continental shelf is the largest carbonate factory in the world; it is three times the length of the Great Barrier Reef. Regrettably, we know little about this massive marine ecosystem. Along with having the largest

Figure 5.35 *The tropical Turtle Beach is a small, headland-bound beach that receives some of the bigger surf on the Arnhem Land coast and is occasionally patrolled by Gove Surf Life Saving Club.* Photo: A.D. Short

temperate seagrass meadows, there is an abundance of carbonate sediment to build the most massive beaches and coastal dunes in the world.

Most Australian beaches are undeveloped, unnamed and found in their natural state, with one-third protected in national parks. As a result, Australia has thousands of pristine, undeveloped beaches with clear water, clean sand and surf, usually bordered by headlands and backed by vegetated coastal dunes.

COASTAL DUNES AND BARRIERS

Coastal dunes

MOST OF AUSTRALIA'S beaches are backed by some form of coastal sand dune system. Occasional strong winds blow the sand inland from the beaches to form the dunes, which are usually stabilised by vegetation (see Figure 6.1). The dunes, however, vary considerably in size, from small, low patches of wind-blown sand to massive dunes rising to heights of 280 metres. Australia has the largest and some of the most extensive coastal dune systems in the world, the longest extending 110 kilometres inland. The location and formation of coastal dunes is closely related to three factors: the beaches that act as the source of sand, the wind that transports the sand inland and the regional climate influencing the wind, dune vegetation and ultimate stabilisation of the dunes.

Australia's coastal dunes occupy an area of 23 500 square kilometres, and back 12 774 kilometres of the coast, or 85 per cent of the sandy coast. They are normally absent from very small, embayed and very low-energy beaches. On average, coastal dunes have a mean elevation of 10 metres, a maximum elevation of 26 metres and extend 1.8 kilometres inland. They contain 280 cubic kilometres (or 280 000 000 000 cubic metres) of marine sand, blown inland from the beaches. This represents an average of 22 000 cubic metres of dune sand for every metre length of beach. The dune systems therefore represent a massive volume of sand, sourced from the marine environment and transported onto the beach by waves, and then blown inland grain by grain, by occasional strong winds, until they eventually

Figure 6.1 *A partly vegetated, incipient foredune, backed by a Casuarina-covered foredune at Salmon Creek, Queensland.* Photo A.D. Short

become stabilised and vegetated (see Figure 6.2). They form a major component of the Australian coastal environment, and some of Australia's major coastal suburbs and resorts are built on dunes, including Bondi and Manly, in New South Wales, the Gold Coast in Queensland and much of coastal Adelaide and Perth.

Dune aerodynamics

Dunes require two conditions for their formation: an abundant source of unconsolidated fine to medium sand and moderate to preferably strong onshore wind velocities. As discussed in Chapter 2, Australian beaches provide an abundance of fine to medium sand, and occasional strong onshore winds are common across the southern half of the continent and along the east coast. All of Australia's largest dune systems are located behind high-energy beaches that are capable of supplying large volumes of sand to the beach, and are exposed to periodic strong to very strong winds, capable of blowing large volumes of sand inland. In contrast, low-energy beaches, wherever they occur, and areas of low to at-most moderate onshore winds, particularly across northern Australia, have limited dune formation.

How wind moves sand

Wind needs to be strong in order to move sand. For most of the time, sand is stationary because wind velocities are too low, which is why on most days you go to the beach it is a pleasant experience. But go to the beach on a windy day, and you will be blasted by flying sand. Wind has to reach a velocity of at least 5 metres per second (18 kilometres per hour) before sand will start moving. This is called the threshold, or 'entrainment velocity', when the wind is strong enough to either move the sand grains along the surface or just lift the sand grains into the air. At 5 metres per second, some fine sand will become airborne; by 6 metres per second (22 kilometres per hour), medium sand will also start moving (see Figure 6.3).

Dune vegetation

Australian coastal dunes support a predictable and fairly narrow range of plant types, as discussed in Chapter 3, although the actual species vary considerably around the coast. The plants can be grouped into pioneer, or primary, species at the front of the dune, then secondary and finally tertiary, or climax, species to the rear. The primary plant types must tolerate the infertile dune sands, exposure to salt spray, strong winds and occasional inundation with salt water, as well as being buried by sand or eroded by waves. As a consequence, only a few species have adapted to this harsh habitat.

The role of dune vegetation

Dune vegetation plays a major role in the shaping, evolution and ultimate stability of coastal dunes, particularly in Australia, where most of the dunes

Figure 6.2 *The coastal barrier system at The Cape Beach, Queensland, is backed by a 1-kilometre-wide, low dune-capped barrier, then a 1-kilometre-wide interbarrier depression that is occupied by a shallow freshwater swamp.* Photo: A.D. Short

Figure 6.3 *Active aeolian sand transport causes regular, small sand ripples, moving from left to right (field of view approximately 2 metres).* Photo: A.D. Short

Figure 6.4 *A hind dune covered in blackbutt forest (Eucalyptus pilularis), with an understorey of bracken (Pteridium esculentum) at Jervis Bay, ACT.* Photo: A.D. Short

Figure 6.5 *A series of weakly developed, horizontal podsol soil horizons (darker layers) and tree roots exposed in the face of an eroded foredune at Big River Cove, Flinders Island in Tasmania.* Photo: A.D. Short

are well to reasonably well vegetated (see Figure 6.4). The most important line of dune vegetation is the row of primary stabilisers – the grasses, creepers and succulents – which grow immediately behind the beach. While these are often sparse in coverage and low in height, they provide sufficient roughness to lower the wind velocity close to the surface, and promote sand deposition and stabilisation (see Figure 6.5). It is virtually impossible for sand grains to pass all the way across a 50-metre-wide, incipient foredune vegetated with primary stabilisers such as *Spinifex* (see Figure 6.6). The sand will be trapped and deposited usually well before it reaches the rear, even under gale-force winds. Vegetation, therefore, is critical in trapping sand immediately behind the beach. If there is no foredune, or if it is breached or insufficiently vegetated, then the sand can be transported further inland.

Dune types

Coastal dunes can be classified into two basic types: regressive and transgressive. Regressive dunes are dunes that build seaward, or regress, on a prograding shoreline. As the beach builds seaward, a series of foredunes paralleling the rear of the beach mark the seaward progression (see Figure 6.7). The foredunes are always vegetated and therefore have stable surfaces and accumulate sand directly off the beach. Transgressive dunes move landward, transgress and bury whatever is behind them. They occur when there is no foredune or there are breaches, or gaps, in the foredune (see Figure 6.8). When active, they are bare and unstable, and can move hundreds of metres to kilometres inland; however, they eventually become vegetated and stabilise. Transgressive dunes derive their sand from directly off the beach and, more importantly, utilise sand eroded from any dunes they move over, reworking pre-existing dunes and re-mobilising their sand.

Regressive dunes

The principal feature of regressive dune systems is the foredune, the foremost or first dune behind the beach, and which forms parallel to the shoreline. They are vegetated and tend to be asymmetrical in shape, higher toward the beach, then sloping inland. They are typically a few metres to as much as 20 metres high, and from a few tens to 100 metres wide. Their shape or morphology is dependent on the vegetation. Regressive dunes are lower and gently undulating, with low, sparse grasses such as *Spinifex*; whereas they are taller and hummocky in taller, dense grasses like the introduced marram (*Ammophila*). The foredune consists of three sections: the incipient foredune, established foredune and the hind dune.

The established foredune lies immediately behind the incipient foredune. Its exposed, seaward side is vegetated by primary stabilisers, grading into low stunted shrubs (see Figure 3.1), while taller shrubs and even some trees grow on its more sheltered leeward side. The dune parallels the beach and usually ranges from a few to several metres in height; Australia's highest foredunes reach up to 20 metres. The foredune is protected from swash and sand burial by the incipient foredune, although it is still exposed to salt spray, strong wind and has poor soils. The strong winds keep the windward vegetation low and often stunted, with plants like the wattle growing as a stunted shrub, and with most plants less than 1 metre high. In lee of the crest and sheltered from the winds, the same plants

Figure 6.6 *An incipient foredune covered in* Spinifex, *located at the base of a higher, shrub-covered foredune, at Lighthouse Beach in New South Wales.*
Photo: A.D. Short

Figure 6.7 *The cleared 5-kilometre-wide foredune ridge plain at Rivoli Bay, South Australia. The plain contains 80 ridges, which have been deposited over the past 5000 years.*
Photo: A.D. Short

Figure 6.8 *Transgressive dunes at Stockton Bight, New South Wales, burying the backing vegetated hind dunes.*
Photo: A.D. Short

Figure 6.9 *The leading edge of a blowout, bordered by vegetated walls at Myall Lakes National Park, New South Wales.* Photo: A.D. Short

Figure 6.10 *A field of nested parabolics at Bald Point in the Northern Territory.* Photo: A.D. Short

grow taller and an understorey of ground cover begins to develop.

Transgressive dunes

Most of Australia's coastal dunes are transgressive dunes – that is, they have been blown or have transgressed inland. Most are now stabilised by vegetation, with only 12 per cent presently actively moving inland (see Figure 6.8). Transgressive dunes cover an area of 2765 square kilometres, mostly across the south and southwest of the continent.

Transgressive dunes and areas of bare blowing sand are part of a natural coastal system and ecosystem, and only in a few instances have been exacerbated by human activity. For the most part, they should be left to evolve naturally, except when they occasionally threaten human habitation or roads.

Blowouts are usually initiated as small depressions or areas of bare sand in the foredune, from which sand is blown inland. This lowers the sand level, forming a depression or deflation basin, and the sand blown inland is deposited in lee of the depression as a raised depositional lobe. The eroding sides of the blowout form erosion rim walls, partly stabilised by vegetation (see Figure 6.9). Blowouts range in size from a few, to hundreds of square metres when fully developed. They then deflate and deepen in the direction of the prevailing onshore wind as they evolve and enlarge, forming an elongate depression. They will continue to erode until they reach the watertable or other cohesive surface, which then forms the base of the deflation basin. Because the base of the basin is close to the water table in southeast Australia, it is usually vegetated by water-tolerant species such as the Knobby Clubrush (*Isolepis nodosa*).

As the blowout elongates in the direction of the dominant onshore wind, it assumes a parabolic shape and becomes known as a 'parabolic dune' (see Figure 6.10). The dune consists of two longer erosional rim walls, or trailing arms: the central elongate deflation basin and the forward-migrating depositional lobe. Parabolic dunes may reach several hundred metres in length, with the depositional lobe climbing to tens of metres in height, while the basin is deflated down to the water table and commonly vegetated by reeds and primary stabilisers.

Some parabolics remain active for centuries, and their depositional lobe can achieve considerable size. As the larger front of the dune extends inland it may widen, leaving in its lee a central deflation basin, with vegetated trailing arms or walls to either side (see Figure 6.11). In areas of strong, uni-directional onshore winds, the parabolic dunes can become quite elongated and when longer than several hundred metres are known as 'long-walled parabolic dunes'. Good examples occur on eastern Cape York Peninsula (shaped by the southeast trade winds), north Tasmania and across southern Australia (westerly winds), and on the central Western Australian coast (southerly winds). The longest dunes reach several kilometres in length (see Figure 6.12). Australia's longest active, long-walled parabolic dune, at 38 kilometres long, is located near Steep Point in Western Australia.

Bare sand dunes

All the dune forms we describe above require vegetation in order to stabilise their sides, so that only the bare leading-edge of the dune moves forward. However, some transgressive dunes are free

Figure 6.11 *A large, 2-kilometre-long, active parabolic dune with bare leading edge and partly vegetated trailing walls, Sensation Beach, South Australia.* Photo: A.D. Short

Figure 6.12 *These long-walled parabolics at Point Louise, in the central west of Western Australia, are maintained by the southerly wind.* Photo: A.D. Short

of vegetation, and when this occurs vegetation plays no role and the dune form is a result of wind action only. In these areas a range of predictable dune types occurs, with their size and form also influenced by the size of the sand grains and the thickness of the dune sand. These bare sand dunes are known as transverse and barchoidal ridge dunes.

Transverse dunes

Transverse dunes are bare sand dunes aligned perpendicular, or transverse, to the prevailing wind. They are asymmetrical in shape, with a lower gradient windward slope (~15 degrees) and a steeper migrating downwind slope (called the 'slip face'), which slopes from 32 degrees in fine sand to 36 degrees in medium sand. Transverse dunes are commonly spaced tens of metres apart and reach several metres in height. In general, the coarser the sand the more widely spaced and higher the dune.

Figure 6.13 *Relatively straight, transverse dunes moving from right to left at Canunda National Park, South Australia. A continuous deflation basin and foredune separate the dunes from the beach.* Photo: A.D. Short

Transverse dunes reach their greatest size at Canunda National Park in South Australia, where some dunes are 20 metres high and spaced 200 metres apart (see Figure 6.13).

Barchoidal dune ridges

The transverse dune ridges tend to be more sinuous, rather than straight. This is because the sand is not thick and some underlying, harder surfaces are exposed in the swale. When this occurs the ridges become sinuous and are called a 'barchoidal ridge'. Transverse and barchoidal dune ridges migrate in the direction of the prevailing wind at average rates between 5–10 metres per year, though higher rates have been recorded during periods of very strong winds.

Sand sheets and sand seas

When transverse dunes occupy a large area (a few square kilometres) they are collectively known as

Figure 6.14 *Extensive areas of bare sand, or sand sheets, at Eucla in Western Australia.* Photo: A.D. Short

a 'sand sheet' or, if very large (a few tens of square kilometres), a 'sand sea'. These sand sheets and seas are rare in coastal locations (see Figure 6.14). Australia's largest areas of bare coastal sand can be found at Dunns Rock (225 square kilometres), Jurien (120 square kilometres) and Black Head (88 square kilometres) in Western Australia, and at The Coorong in South Australia (116 square kilometres).

As transgressive dunes move inland they usually leave behind an extensive area of deflated sand, called a 'deflation basin'. The basins may parallel the rear of the beach for several kilometres, lying between the rear of the foredune and inner edge of the transgressing dunes. The surface of the basin usually lies close to the watertable and consists of bare sand, hard surfaces, low active dunes and vegetated patches, resulting in a low-amplitude, hummocky surface that is vegetated by primary and secondary species, as well as freshwater reeds in some low areas close to the watertable. The active dunes continue to feed sand to the transgressive dune system.

Clifftop dunes

An unusual, but nonetheless common, transgressive dune form occurs on top of, or in lee of, sea cliffs up to 200 metres high. Clifftop dunes form where high-energy waves provide an abundant supply of sand to build beaches at the base of the cliffs. Very strong winds are then required to build a steep sand ramp up the face of the cliffs (see Figure 6.15), blow the sand up the ramp and over the cliffs, then blow the sand in places up to 20 to 30 kilometres inland, usually as parabolic and long-walled parabolic dunes. Most of the beaches and sand ramps that supplied the dunes around the Australian coast have since been

Figure 6.15 *Active sand ramps and clifftop dunes at Cape Tournefort in South Australia.* Photo: A.D. Short

Figure 6.16 *Partly active, nested-parabolic clifftop dunes at Avoid Bay in South Australia. These dunes are cut off from their sand supply by 40-metre-high, calcarenite cliffs.* Photo: A.D. Short

eroded, leaving the dunes stranded on the clifftops and usually stabilised by vegetation and separated from the cliff edge by a bare, deflated rocky surface (see Figure 6.16).

Where the clifftop dunes have been dated, they range in age from 12 000 to 5000 years old, indicating that most were forming as the sea level was below its present level and still rising. Some continued to form once the sea level stabilised around 6500 years ago. However, in most cases the supply of sand was rapidly exhausted after the sea level stabilised, and the sand ramps – and in most cases the beaches – were eroded, with only the dune left as evidence of this former activity. Most of the beaches and ramps were rapidly eroded because, just as the high wave energy was able to supply vast quantities of sand rapidly to the shore to help build the dunes, the same wave energy just as rapidly depleted and exhausted the finite supply, and was then able to erode the beaches.

Coastal barrier systems

Coastal barriers represent the long-term accumulation of wave, wind and tide-deposited marine sediments along the shore. They can range in size from low, narrow cheniers and beach ridges, to massive, high-energy beaches and sand dunes the size of Fraser Island. There are 2500 barrier systems around the Australian coast, which occupy 12 000 kilometres of the coast. As such, they are the most dominant landform around the coast. They average 5 kilometres in length, with one barrier often fed by several adjoining beaches. They extend on average 2 kilometres inland, with an average height of 10 metres but with a considerable range in size and volume. Barriers can be divided into the following categories:

- regressive, where the shoreline is building seawards
- stable, where it is stationary
- transgressive, where there is movement of dunes inland
- inherited, where Pleistocene barriers exert a control on present coastal morphology.

Regressive barriers

Regressive barriers, like regressive dunes, are beach and backing barrier systems that are building seaward where the shoreline is regressing. They occur in areas where sand is being delivered to build the beaches and backing dunes. In wave-dominated environments, regressive barriers usually consist of a series of foredune ridges; that is, a sequence of shore-parallel foredunes. The most extensive regressive (foredune) barriers occur along the eastern Gulf of Carpentaria. Along the 900 kilometres of coast between the Northern Territory border and Worbody Point, near Weipa, there are 26 low-energy, regressive barrier systems totalling 2400 square kilometres in area. They also occur as individual systems along the east Queensland coast (for example, Cowley Beach); along the New South Wales coast (Tuncurry, Ocean Beach–Woy Woy, Seven Mile Beach, Broulee–Moruya (see box 'Broulee–Moruya regressive barrier' and Figure 6.17, below), Merimbula and Disaster Bay); Victoria (Ninety Mile Beach); Tasmania (Nine Mile and Seven Mile beaches), South Australia (Rivoli (see Figure 6.7), Guichen and Lacapede bays), Western

BROULEE–MORUYA REGRESSIVE BARRIER

The Broulee–Moruya barrier system is one of the largest and best-studied regressive barriers on the southeast coast. It consists of a series of 40 inner-beach to outer-foredune ridges that were deposited between 6000 and 2000 years ago, as the shoreline prograded 2 kilometres seaward. Note in Figure 6.17 how sand rapidly infilled the inner shallower part of the bay, resulting in rapid shoreline progradation of over 1 kilometre in 1000 years, then slowed as more sand was required to fill the outer, deeper part of the embayment and finally stabilising as the sand supply was apparently depleted. Such a scenario of initial rapid infill, followed by a slowing and cessation of sand supply, is typical of the regressive barriers that have been studied around the southeast coast.

Australia (Safety Bay and Eighty Mile Beach) and more exposed parts of the Arnhem Land coast.

Cheniers

Cheniers are low ridges, usually of shell-rich sediment, deposited by waves on a high-tide mud flat. Their height is limited to the reach of waves, usually just above high-water level, and their width up to several tens of metres. They extend alongshore, usually in a relatively straight fashion, for hundreds of metres to a few kilometres, often terminating in recurved ridges at creek mouths.

Cheniers occur across northern Australia in sheltered, mud-rich tidal flat environments, commonly in association with the tide-dominated deltas described in Chapter 4. These are very low-energy, tide-dominated systems that are periodically

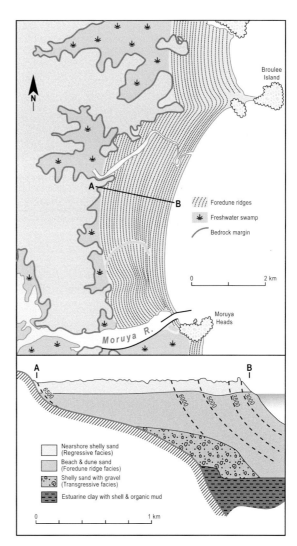

Figure 6.17 *The Broulee–Moruya foredune ridge plain is an 8-kilometre-long, 2–3-kilometre-wide regressive barrier containing up to 40 ridges, which were deposited between 6500 and 3000 years ago. The cross-section (A–B) shows a basal layer of estuarine sediments, which were deposited at a lower stand of the sea, between 9000 and 10 000 years ago.*

Figure 6.18 *Cheniers deposited on wide intertidal mud flats at Point Parker, southern Gulf of Carpentaria.*
Photo: A.D. Short

Figure 6.19 *A closely spaced series of more than 10 beach ridges at Pormpuraaw, western Cape York Peninsula.*
Photo: A.D. Short

exposed to wave activity. The waves wash over, erode and winnow shells and coarser material from the mud flats, and then transport them shoreward to be deposited at the limit of wave attack, as a low ridge of shell-rich sand. This may occur in lee of an intertidal mangrove fringe, leaving the chenier fronted by a band of mangroves. When exposed to tropical cyclones and accompanying higher waves and storm surge, the mangroves may be killed and the cheniers undergo accelerated landward migration. Once formed, their crests become vegetated by *Casuarina*, and dense vine-thicket in the Northern Territory. If the shoreline progrades a series of cheniers may develop, located tens to hundreds of metres apart, with each chenier separated from the other by intertidal mud flats.

Across the northern coasts where a slight fall in sea level has occurred some series of cheniers extend across tens of kilometres of coastal plain, such as in southern van Diemen Gulf, along the southern Gulf of Carpentaria, in Princess Charlotte and Cape Bowling Green bays and Broad Sound. Cheniers occur across the north of Australia and along sheltered parts of the Pilbara coast and in King Sound, along sheltered bays on the Kimberley coast, in association with the many river mouths that reach the Northern Territory coast and on the eastern Cape York Peninsula–Queensland coast (see Figure 6.18). They have only been found in southern Australia at Cullendulla Creek in Batemans Bay (New South Wales) and Davenport Creek in western South Australia.

Beach ridges

Beach ridges are shore-parallel ridges of wave-deposited sediment, formed on intertidal sand flats, with heights usually less than 3 metres, and whose swales are usually intertidal. They differ from cheniers, in that cheniers lie on top of a mud flat, whereas beach ridges are on top of a sand flat, and foredunes are wind-deposited ridges sitting atop beach deposits. In Australia, the low-energy ridges commonly form in lee of wide sand flats fronted by seagrass meadows, from which carbonate-rich shelly sediment is winnowed from the fronting sand flats and meadows to be deposited on the ridges. As with cheniers, a prograding shoreline will form a series of beach ridges usually separated by intertidal sand flats, with low dune vegetation on the ridges and samphire or mangrove vegetation in the swales (see Figure 6.19).

Stable barriers

Stable barriers usually consist of a low-energy, reflective beach and a backing foredune or beach ridge that have remained essentially unchanged for a considerable period, usually a few thousand years. While the beach may undergo occasional erosion and accretion of the shoreline, the backing dune system remains largely untouched and stable. Stable barriers tend to occur in small, sheltered, headland-bound coastal embayments, often facing away from the dominant waves and winds. Because of their embayed location they have a stable and fixed sediment budget, and because they are sheltered they are not exposed to high waves and strong winds. This results in a low-energy beach and sheltered dune, the latter often allowing climax vegetation in well-developed soil to grow close to, and even to overhang, the beach.

Transgressive barriers

Transgressive barriers are barriers that are moving, or transgressing, landward. The movement may include the shoreline and/or coastal dunes, resulting in one of two types: a transgressive-shoreline barrier, where the shoreline is actively eroding (which may or may not have transgressing dunes), or a transgressive dune-barrier, where the shoreline may be stable but the backing dunes migrate inland.

Transgressive-shoreline barriers

Transgressive-shoreline barriers occur in areas that have a negative sediment budget, where more sediment is being lost out of the system than is moving in. Sediment is usually lost to longshore sand transportation, but can also result from loss onshore to dunes or into tidal inlets, or offshore during big seas. The best examples of loss to longshore sand transportation occur along the northern New South Wales and southeast Queensland coast. The result is retreat of the shoreline and, in places like Broadwater National Park (New South Wales), Rainbow Beach and Fraser Island (Queensland) exposure and erosion of older Pleistocene dunes (see Figure 6.20). The eroded Pleistocene dunes are normally bleached white in colour, may contain multiple old-soil horizons, commonly called 'coloured sand', and may be underlain by a dark, indurated sand layer, known as 'coffee rock', due to its resemblance to compacted coffee grounds.

Figure 6.20 *Eroding beach and dunes along Seventy-Five Mile Beach, Fraser Island, Queensland. The irregular dark protrusions are coffee rock.* Photo: A.D. Short

Transgressive dune barriers

Transgressive dune barriers are the most extensive of the Australian barrier systems, particularly as they are often large in size. They include several massive, trade-wind-driven dune systems on:

- eastern Cape York Peninsula, between Sharp Point and Cape Bedford
- Ramsay Bay (Hinchinbrook Island)
- Port Clinton
- Nine Mile Beach (Byfield National Park), Curtis Island
- the massive southeast Queensland sand islands
- in New South Wales, at Killicks Beach, the Myall Lakes region, Stockton Bight, Kurnell Peninsula and Bhewherre Beach
- in Victoria, along the Croajingolong National Park coast and in Discovery Bay

- along parts of Tasmania's west and eastern north coast
- in South Australia, along the Canunda National Park coast, much of the Coorong (Sir Younghusband Peninsula), the south and west coast of Kangaroo Island, the southern tip of Yorke Peninsula and along much of the western Eyre Peninsula coast.

In Western Australia there are massive transgressive dunes in the Bight, also in parts of the Cape Arid and La Grande National Parks, and spread along much of the south, southwest and central west coast, as far north as Coral Bay (see Figure 6.21). Across northern Australia, transgressive dunes occur on only a few shorelines that are well exposed to the southeast trades, with only one system on the east Kimberley coast at Cape St Lambert; and in the Northern Territory on the exposed shore east of Cobourg Peninsula and south of Cape Arnhem, and on the eastern shore of large islands, such as Groote Eylandt.

Distribution

Coastal barriers are distributed around the Australian coast in three broad provinces:

1. The 11 987-kilometre-long eastern province, located between western Cape York Peninsula and eastern Victoria and Tasmania, is the longest, with 45 per cent of the coast backed by 1150 barriers, of which 9 per cent consist of active transgressive dunes. They have a total volume of 96.5 cubic kilometres and an average volume of 21 667 cubic metres per metre of beach. This province includes the massive southeast Queensland sand islands.

2. The southern–western province includes the west coasts of Tasmania and Victoria, all the southern Australian coast and Western Australian coast as far north as North West Cape. There are only 589 barriers, but they back 48 per cent of this 9587-kilometre-long shoreline and attain the greatest volume (175 cubic kilometres) and volume-per-metre of beach (42 157 cubic metres per metre). These are also Australia's least stable barriers, with 17 per cent consisting of bare sand dunes, except in South Australia, where as much as 35 per cent is unstable. These most extensive and least stable barriers have been produced and maintained by the high waves and winds of the southern–southwestern coast.

3. The equally long northwest province, which includes the Pilbara, Kimberley and western Northern Territory, has more barriers (737), but these back only 22 per cent of the coast and contain only 10 cubic kilometres of sand, deposited at a rate of 4161 cubic metres per metre of beach. The low level of barrier development is due directly to the low levels of wave and wind energy, which supply lower rates of sand and maintain more stable barriers.

Inner barriers

Around most of the Australian coast, previous sea-level highstands have left a wide range of Pleistocene coastal deposits, usually consisting of former beach-dune systems (see Figure 6.22). Barriers tend to be preserved on lower-gradient sections of coast, particularly where the coast has

been uplifted. The particularly extensive sequence of barriers and intervening depressions marking successive interglacial highstands in the Murravian Gulf, extending 500 kilometres inland, have already been described in Chapter 1 (see Figure 1.8).

Buried Pleistocene barriers

Buried Pleistocene barriers tend to occur on steeper and more exposed, higher-energy shores. As sea level reached its present level, the modern beach abutted the older Pleistocene system and proceeded to blanket the older barrier. In other, more exposed locations, the earlier deposits have been buried or blanketed by each successive layer of highstand dune sand, massive layered sand dunes forming in places up to 200 metres high. Each layer is capped by a darker soil horizon, with a basal indurated zone (coffee rock, or palaeosol) close to sea level – the total exposure is commonly referred to as 'coloured sands'.

Figure 6.21 *The linear ridges to the right are deflated foredune ridges that have been overrun by the transgressive dunes to the left, indicating an initial period of ridge development and shoreline growth, followed by the transgression, Point Edgar, Western Australia.*
Photo: A.D. Short

These include parts of the northern New South Wales coast, the southeast Queensland sand islands, including Fraser Island, and the calcarenite (lithified sand dunes) cliffs spread across the south coast and along the Zuytdorp Cliffs of Western Australia, as well as some exposed locations in Arnhem Land.

Figure 6.22 a) *Inner Pleistocene foredune ridges, behind the modern beach and foredune at Diamond Head, New South Wales;* **b)** *Looking north across an eroded Pleistocene inner barrier (foreground), 3-kilometre-wide salt flats and the outer Holocene barrier, southern Gulf of Carpentaria, near Northern Territory–Queensland border.* Photos: A.D. Short

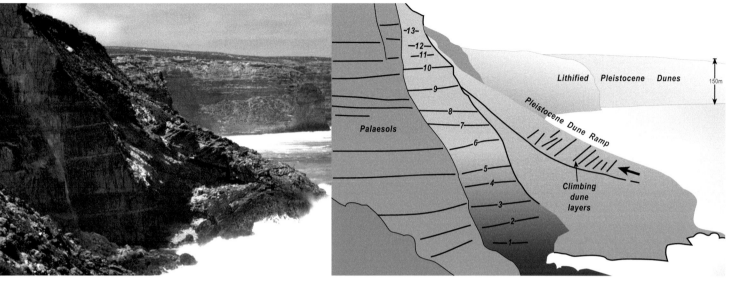

Figure 6.23 a) *150-metre-high, calcarenite cliffs, formed from lithified sand dunes at Groper Bay on Eyre Peninsula, South Australia.* Photo: A.D. Short

b) *Each of the 13 bands represents a preserved soil horizon (palaeosol) formed on a layer of dune sand, each layer deposited approximately 100 000 years apart at sea-level highstands. The lowest layers are at least 2.5 million years old. A lithified sand ramp overlays the palaeosols, with the layers of the sloping ramp preserving the incline of the once-climbing sand dune.*

Dunerock (calcarenite)

In many places, the Pleistocene barrier sands have been partially cemented into calcarenite. Where these barrier sands consist of dunes, the calcarenite is also called 'aeolianite', commonly known as dunerock. This occurs along the southern Australian coast, from Darby River at Wilson's Promontory, across western Victoria and much of South Australia (see Figure 6.23), including the entire south and west coast of Kangaroo Island, along the Great Australian Bight and much of the south-central coast of Western Australia, as well as in scattered locations around the Kimberley and Northern Territory coast, including on Groote Eylandt. Calcarenite occurs when a sand dune is lithified into a rock. This is

caused by the formation of calcrete soils in semi-arid to arid, carbonate-rich dune and beach sands. In these environments, sufficient precipitation and humic acids occur to dissolve calcium carbonate in the upper soil, which then seeps into the subsurface, but not enough to flush it out of the system. As a result, the carbonate-rich groundwater evaporates and precipitates the carbonate between the sand grains, thereby gradually cementing them together. Over tens of thousands of years, the dunes gradually lithify and become a weakly consolidated dunerock.

Figure 6.24 *The shore-parallel beachrock reef, with a strong rip running out through the gap at Anxious Bay, western Eyre Peninsula in South Australia.* Photo: A.D. Short

Figure 6.25 *A regressive-recurved beach-ridge barrier at the mouth of Accident Creek, southern Gulf of Carpentaria.* Photo: A.D. Short

BEACHROCK

Another common form of calcarenite is a lithified beach commonly known as 'beachrock'. This is more prevalent in the tropics, where the combinations of carbonate-rich beach sediment and warm water result in a supersaturation of calcium carbonate in the seawater in the intertidal zone of the beach. This leads to development of calcium carbonate that cements the sand grains. Beachrock can form in a few years, embedding many modern artefacts including jetsam such as drink bottles and cans. Beachrock is also common along parts of the carbonate-rich southern Australian coast, reaching its best development in Anxious Bay, near Head of Bight (see Figure 6.24) and at Shoal Point in south Western Australia. When the overlying beach is eroded the beachrock remains as a shore parallel reef, causing heavy wave breaking on the reefs and lower-energy lagoonal conditions to the lee. However, gaps in the reefs are usually the focus of strong topographically controlled rip currents.

Australian dune-barrier systems

In this section we review the dune-barrier systems in each state, highlighting some of the major systems in Australia's 2476 individual coastal dune-barrier systems. Our review begins in Queensland, working clockwise around the coast.

Queensland

Queensland has 6680 kilometres of coastline, of which 46 per cent (3103 kilometres) is backed by 493 barrier systems. These barriers represent 12 per cent of the total number of barrier systems by area, but 26 per cent by volume – the volume largely contained in the massive transgressive dunes on eastern Cape York Peninsula and particularly the southeast sand islands. The major Queensland systems include the 913 kilometres of generally low-energy and low regressive chenier, beach ridge and foredune ridge barriers, which occupy 62 per cent of the southern Gulf, particularly the western Cape York Peninsula coast. These systems are supplied with sediment from the numerous river systems (discussed in Chapter 4), which have prograded the coast and built 900 kilometres of generally low but stable cheniers in the southern gulf, and beach and foredune ridges along the western cape shore, with most of the barriers bounded by river and creek mouths (see Figure 6.25).

The eastern coast of Cape York Peninsula, south to Cooktown, faces into the seasonally strong southeast trade winds. While the southeast waves are reduced by the Great Barrier Reef, the strong winds – particularly in winter – have blown sand from the most exposed, southeast-facing beaches to form some particularly extensive transgressive dune systems. In all, 91 barriers occupy 56 per cent of the 1032 kilometres of shore. Queensland's largest mainland transgressive dune systems are located on eastern Cape York Peninsula, at Cape Grenville and Cape Flattery. Both capes protrude east, with the beaches on their southern side facing directly into the southeast trade winds. These persistent winds have generated the waves to move the sand onshore and alongshore to the relatively higher-energy, tide-modified beach systems; then the same winds have blown sand from the beaches as long-walled parabolic dunes, up to 28 kilometres inland at Cape

Figure 6.26 *Trade wind-aligned, transgressive dune fields at False Orford Ness, eastern Cape York Peninsula.*
Photo: A.D. Short

Flattery (see Figure 6.26) and 29 kilometres at Cape Grenville. Preliminary dating indicates that the Holocene dunes were initiated between 8000 and 7500 years ago. The dunes average 30–40 metres in height, reaching 140 metres in lee of Cape Flattery. Because of the uni-directional southeast winds and the dense tropical vegetation cover, the parabolic dunes are both well-developed and long, with many individual dunes between 10 and 15 kilometres in length – the most extensive and longest longwalled parabolic field in Australia. There are several smaller systems of a similar nature along the east Queensland coast, scattered between Sharp Point and Nine Mile Beach.

The 3325-kilometre-long east Queensland coast between Cooktown and Hervey Bay has 310 barrier systems, which occupy 36 per cent of the coast, the lowest percentage for the Queensland

SOUTHEAST QUEENSLAND SAND ISLANDS

The largest coastal dune systems in the world are located along the 360-kilometre-long southeast Queensland coast, between Fraser and South Stradbroke islands. Fraser Island dominates in size and volume (see Table 6.1) – in fact, it is by far the largest accumulation of coastal sand on the planet. The sand islands have a similar mode of formation to the east-coast dune fields. They formed during sea-level highstands over the past 2 million years, with each successive highstand adding another layer to the islands. The highest dunes on Fraser, Cooloola, Moreton and North Stradbroke islands average 60–100 metres in height, which means each successive highstand adds on average about 5 metres of sand.

Sand island	Area km²	Mean height m	Maximum height m	Volume km³
Fraser	1840	100	244	184
Cooloola	600	60	237	36
Bribie	180	5	10	1
Moreton	150	100	280	15
North Stradbroke	260	100	218	26
South Stradbroke	195	10	20	2
Total	3225			264

Table 6.1 *Area and volume of southeast Queensland sand islands*

Figure 6.27 a) The Don River delta, with active sand spits migrating north from the river mouth. The river delivers large volumes of sand, which is reworked northward by trade wind-generated waves. b) The 3.5-kilometre-wide Cowley Beach foredune ridge plain, in Queensland, contains 60 ridges covered in dense tropical vegetation. Photos: A.D. Short

coast. This is due to the number of large, mangrove-lined bays and extensive sections of rocky shore. The barriers include low-energy cheniers and beach ridges, with the most extensive occuring in association with several larger regressive river deltas, including the Don River (see Figure 6.27a) and the large, regressive beach-foredune ridge systems at Cowley Beach (see Figure 6.27b).

Sand has accumulated in southeast Queensland for four reasons. There is a source of plentiful fine sand. The sand originates in the granite of the New England Fold Belt, which has been eroded and transported to the coast by the northern New South Wales river systems, particularly during lower stands of sea level. During sea-level rise and

Figure 6.28 Scarped, coloured Pleistocene dunes at Rainbow Beach, Queensland.
Photo: A.D. Short

highstands, the prevailing southerly waves move this sand onshore and northwards along the coast. The southeast trade winds have blown the sand from the beaches to form the massive dunes. At the border there is an inflection in the coast, enabling the sand to be deposited seaward of the shore on the inner continental shelf, with each of the high islands having a bedrock high point that anchors the northern tip of each island. These are Point Lookout for North Stradbroke Island, Cape Moreton on Moreton Island, Double Island Point at Cooloola, and Indian Head and Waddy Point on Fraser Island.

Each of the high islands therefore consists of successive layers of dune sand, supplied to the higher-energy beaches by waves and rising sea level, and then deposited as parabolic dunes atop the existing sand dunes. Over time the layers built the islands higher and higher, as well as wider. The older dunes are

buried along with their soils (palaeosols), which when exposed at the coast are better known as 'the coloured sands of Fraser Island and Cooloola', with the basal coffee rock exposed along the shore (see Figure 6.28).

New South Wales

The coast changes dramatically at the Queensland–New South Wales border, as the large sand islands give way to smaller embayed beaches and barriers. The entire New South Wales coast is dominated by rocky headlands, with intervening beaches and barriers, and 133 estuarine systems. A total of 755 beaches and 278 barriers occupy 58 per cent of the 1590-kilometre-long coast. While New South Wales has 6.3 per cent of the Australian coast, it contains only 3.5 per cent of Australia's barrier by area and 2.8 per cent by volume.

THE MYALL LAKES AND STOCKTON BIGHT

The largest coastal dune and barrier system in New South Wales occupies two sections of a 90-kilometre-long stretch of coast between Seal Rocks and the mouth of the Hunter River at Newcastle. On the north side of Port Stephens is the 42-kilometre-long Myall Lakes system and its 99 square kilometres of coastal sand dunes, while between the Hunter River mouth and Buribi Point is the 70-square-kilometre Stockton Bight dune system (see Figure 6.29), which is similar in size to the smaller Queensland sand islands. The New South Wales systems have similarities but also important differences to the Queensland systems. First, like the Queensland sand islands, these large barriers have formed here because they lie downdrift of a major sand source, the Hunter River. The energetic southerly waves have moved the sand northward at various sea levels, then moved it onshore during rising and contemporary sea levels. Second, it has accumulated in this region because, unlike the southeast Queensland coast, the Hunter coast trends east, with the bedrock valleys trending north–south and forming a natural series of sand traps, between the Hunter and Buribi Point, and between Port Stephens and Seal Rocks–Smith Lake. Third, the eastward trend also increasingly orientates the long beaches to the south, into the high southerly waves and strong southerly winds, both of which are required to supply the sand to the high-energy beaches, then blow it inland as massive dunes. Between Big Gibber and Seal Rocks the modern dunes blanket the previous Pleistocene dune system, producing the largest dunes in New South Wales – extending up to 6 kilometres inland and rising in places to 160 metres.

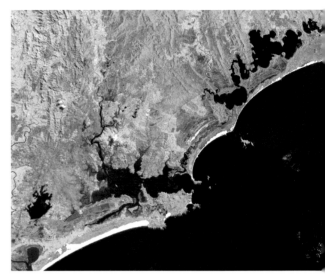

Figure 6.29 *A satellite image centred on Port Stephens, a drowned river valley, with the Myall Lakes barrier systems extending 40 kilometres to the north to Seal Rocks, and the Stockton Bight barrier extending 30 kilometres south to the mouth of the Hunter River at Newcastle.* Image: © Commonwealth of Australia, ACRES, Geoscience Australia

Victoria

Victoria has two barrier coasts, the 440-kilometre-long coast to the east of Wilson's Promontory that faces into Bass Strait, and the 744-kilometre central and west coast that is exposed to the Southern Ocean. The east coast has 49 barriers that occupy 93 per cent of the coast, much of this taken up by the 222-kilometre-long Ninety Mile Beach–Gippsland Lakes system, whose barrier occupies an area of 660 square kilometres. This is a long, but relatively narrow, regressive (foredune ridge)

barrier, now eroding in places. Further east, along the Croajingolong coast, are a number of more extensive transgressive barrier systems between Point Ricardo and Rame Head, the latter capped by well-vegetated clifftop dunes. These culminate in the 200 square-kilometre Mallacoota–Cape Howe systems (see Figure 6.31), the largest transgressive system in the state and which spills over into New South Wales (see Figure 1.13).

The western Victorian coast has 337 kilometres of barrier coast, containing 85 barriers that together occupy 43 per cent of the coast. The largest barriers comprise transgressive dune systems and include Darby River, Venus Bay, Mornington Peninsula, Station Beach–Cape Otway, Flaxman Hill, Bridgewater Bay and Discovery Bay. Discovery

Figure 6.30 a) *The southern tide-dominated section in Corner Inlet breaks the Ninety Mile Beach into a series of low, wave-washed barrier islands separated by large tidal inlets.* **b)** *The northern Bunga Arm section, near Lakes Entrance, is eroding slowly, narrowing and almost breaching the barrier.* Photos: A.D. Short

Bay is Victoria's westernmost and highest-energy beach and barrier system.

Tasmania

Tasmania's barriers can be divided into two regions: the more moderate-energy north and east coast, and the high-energy west coast, including King Island. The 448-kilometre-long north coast is exposed to strong westerly winds, which tend to blow offshore and alongshore in the west. East of Launceston these winds begin to form extensive transgressive dunes in lee of west-facing shores. The result is a lower-energy, more regressive western section and higher-energy eastern section, with a total of 168 barriers occupying 50 per cent of the coast. Regressive

NINETY MILE BEACH

Ninety Mile Beach is one of Australia's longest beach systems and in its own way is a rather unusual Australian beach and barrier system. Like the three other long Australian beaches, Ninety Mile Beach is located within a sedimentary basin, which lacks the rocky headlands that dominate other coasts. This is why the beach is so long. The prevailing westerly gales in eastern Bass Strait result in longshore sand transportation to the northeast.

Ninety Mile Beach commences in the south as a series of low, tide-dominated sand islands. Waves are low, tidal range is up to 3 metres and the prevailing westerlies blow offshore. To the north of the islands, the meso-tidal and low waves permit a series of five large tidal inlets to drain the 350-square-kilometre Corner Inlet; the inlets having substantial ebb and flood tidal deltas separated by low, bare, unstable barrier islands that are devoid of dunes (see Figure 6.30a). Only as the waves begin to increase to about 1 metre in height – by McLaughlin Beach – does the beach become continuous for 127 kilometres to Lakes Entrance, its longest unbroken section. The southern section near Seaspray is eroding, with lesser erosion along much of the beach up to Lakes Entrance (see Figure 6.30b). It then continues for another 8 kilometres to Red Bluff, a slight bedrock inflection, before continuing on for 25 kilometres to its northern boundary at the mouth of the once-mighty Snowy River. The total length of beach between the Snowy River and Snake Island is 222 kilometres (128 miles).

Figure 6.31 *Active blowouts and parabolic dunes north of Mallacoota, with Lake Barracoota and Howe Hill in background.* Photo: A.D. Short

Figure 6.32 *A series of long-walled parabolics in lee of the high-energy Waterhouse Beach, north Tasmania.* Photo: A.D. Short

barriers capped by foredune ridges is most prominent along the western section, with the largest systems at Perkins Island and Anthony Beach, where up to 50 beach-foredune ridges are present, and in the east at Badgers and Green beaches, which both have 25 foredune ridges. To the east, the barriers become increasingly transgressive, with the largest systems at Double Sandy Point, Waterhouse Beach and Boobyalla Beach, where long-walled parabolic dunes extend a few kilometres inland (see Figure 6.32).

The 1097-kilometre-long east coast is a headland-dominated shore with 559 generally small, embayed beaches and 68 barrier systems. The westerly winds predominantly blow offshore, with most of the barriers low to moderate in height and consisting of a few stable foredune ridges (see Figure 6.33). The most extensive regressive barriers are Nine Mile Beach at the top of Great Oyster Bay, Reeban Spit and Hobart's Seven Mile Beach. The only transgressive dune system is the south-facing Peron Dunes.

The west coast consists of three parts. The 448-kilometre-long section south of Macquarie Harbour, which is dominated by rocky coast and generally small-embayed barriers, has 226 beaches but only 50 barriers occupying 59 kilometres of coast. They range from sheltered, stable, single-foredune systems, to several moderate-sized transgressive systems, the largest at Towterer Bay and Stephens Bay (see Figure 6.34), where the dunes extend 2 kilometres inland. The northern half of the coast between Macquarie Harbour and Cape Woolnorth has a similar length (244 kilometres), but three times as many beaches (174). While the north has only 27 barriers, they occupy 119 kilometres (50 per cent) of the coast and include some extensive transgressive dune systems. Immediately north of the harbour

Figure 6.33 *The headland-bound, lower-energy beaches at Little Musselroe Bay are typical of the east Tasmanian coast.* Photo: A.D Short

Figure 6.34 *An active blowout-parabolic dune, burying dense vegetation at Stephens Bay, Tasmania.* Photo: A.D. Short

mouth, at Hells Gate, is a series of more extensive transgressive barrier systems, the largest being the 32-kilometre-long Ocean Beach that extends up to 7 kilometres inland. Other larger systems are at Four Mile Beach, Lagoon River, Sandy Cape and Arthur Beach. There is only one small regressive system in the sheltered Temma–Polly Bay region. A third section is located on King Island, whose 100-kilometre-long western shore is exposed to the full force of the waves and winds of the Southern Ocean. The waves have delivered primarily calcareous marine sands (80–90 per cent carbonate) to the high-energy beaches, which have then been blown inland up to 4 kilometres and overlie older and more extensive Pleistocene dunes.

South Australia

The open coast of South Australia, including Kangaroo Island, contains 212 barrier systems that occupy 1580 kilometres (45 per cent) of the coast. These systems cover an area of 2436 square kilometres, or 10.3 per cent by area and 14 per cent by volume of all Australian barrier systems. This coast has the most unstable dunes systems in Australia, a product of the aridity as well as the high levels of wave and wind energy. The coast contains some of the more dramatic coastal transgressive dunes, including more than 900 kilometres of Holocene clifftop dunes, most sitting on top of Pleistocene calcarenite cliffs.

Western Australia

Western Australia occupies 30 per cent of the Australian coast and contains 31 per cent of the barrier by length, 34 per cent by area and 46 per

167

Figure 6.35 a) Deflated surfaces (foreground) and transgressive dunes at Dunns Rock; *b)* The foredune ridge plain at Windy Harbour, south coast of Western Australia. Photos: A.D. Short

AUSTRALIA'S MOST EXTENSIVE COASTAL DUNE SYSTEM – KANIAAL DUNES

Australia's most extensive Holocene dune system extends inland into the Nullarbor from Kaniaal Beach. The beach commences at Twilight Cove, and forms the first section of a series of low-energy sandy beaches that continue 300 kilometres east to the South Australian border. Along Kaniaal Beach, the dunes back the first 40 kilometres of beach. They trend to the east–northeast at an angle to the shore and have blown up to 120 kilometres inland (see Figure 6.36). The dunes have been deposited as predominantly nested, long-walled parabolics during at least three episodes of dune activity, probably in the early to mid-Holocene period. They are relatively low, with an average height of 20 metres and maximum of 32 metres. Today, only 2 per cent of the dunes are unstable and active, while the rest is vegetated with dense shrubs and mallees of the Roe Plain. They have an area of 1000 square kilometres, and represent approximately 10 cubic kilometres of sand.

cent by volume. The reason for this dominance is the extensive south coast that is exposed to the higher-energy Southern Ocean waves. These have delivered large quantities of sand to the shore, and strong west-to-south winds have blown the sand inland to form Australia's largest barrier systems. The south coast has a total barrier area of 4681 square kilometres, while the south west coast has 2123 square kilometres, the Pilbara 936 square kilometres and the long, although sheltered Kimberley coast only 223 square kilometres.

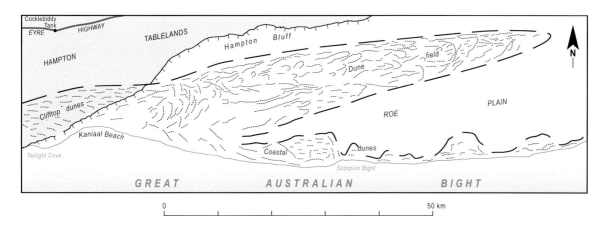

Black Head to Black Point

Western Australia's highest-energy beach is the Meerup–Warren–Yeagarup–Jaspar beach system, which extends essentially unbroken for 50 kilometres between Black Point and Black Head. It is backed by a massive 380-square-kilometre dune system (see Figure 6.37). The dunes average 100 metres in height, reaching well over 200 metres in places and resulting in approximately 38 cubic kilometres of dune sand. The probable source of most of the predominantly quartz sand (80–95 per cent quartz) is the granite that dominates the river catchments.

The west coast

The west coast, from Geographe Bay north to Safety Bay, is dominated by stable barriers with high foredunes and regressive barriers. The most extensive foredune ridge plains in Australia are located between Mandurah and Warnbro–Kwinana, where there are up to 90 foredune ridges covering 110 square kilometres. North of Mullaloo the strong southerly sea breezes generate transgressive barriers with north-trending, parabolic to long-walled

Figure 6.36 The Kaniaal dunes commence at Twilight Cove and extend to the east for 110 kilometres, covering an area of 1000 square kilometres.

Figure 6.37 The high-energy, rip-dominated Warren Beach, backed by a 500-metre-wide deflated dune surface, then the beginning of the massive, partly vegetated, transgressive dune systems. Photo: A.D. Short

WARNBRO, LANCELIN AND THE PINNACLES

Warnbro Sound is a 7-kilometre-wide, west-facing, semi-circular embayment, with a curving, 11-kilometre-long shoreline. Over the past 6000 years sand has moved into the bay and the shoreline has prograded seaward as much as 7.5 kilometres, through the development of up to 90 shore-parallel foredune ridges (see Figures 6.38 and 6.39). The ridges extend south of Becher Point as far as Mandurah and north of Mersey Point into Cockburn Sound. The entire system represents one of the largest areas of Holocene shoreline progradation on the Australian coast. Today, however, the ridges are rapidly being covered by housing estates.

In contrast, north-trending transgressive dunes dominate most of the 350 kilometres of coast between Quinns Rocks and Cape Burny. The best-known dune systems are at Lancelin, where they extend 5–15 kilometres inland, and along the coast for 17 kilometres, occupying 170 square kilometres (see Figures 6.39 and 6.40). The 30 square kilometres of bare sand attract offroad drivers to the pristine dunes. The most visited dune field, however, is at the Pinnacles, near Cervantes. Here, calcarenite again plays a major role, forming a domed cast around the root system of former trees. Subsequent dune transgression has deflated the surrounding unconsolidated dune sand and surface, exposing the once-buried casts as prominent 'pinnacles' (see Figure 6.40).

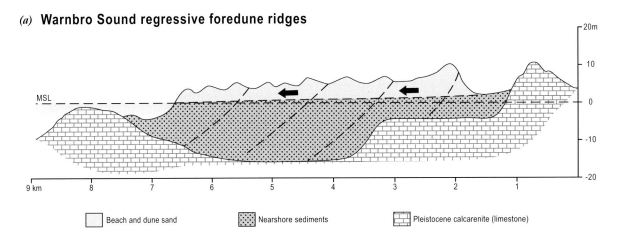

(a) **Warnbro Sound regressive foredune ridges**

Beach and dune sand   Nearshore sediments   Pleistocene calcarenite (limestone)

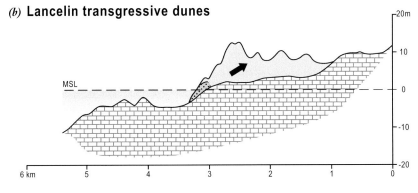

(b) **Lancelin transgressive dunes**

Figure 6.38 a) *Cross-section of Warnbro Sound's foredune ridge plain near Rockingham, with the Pleistocene calcarenite outcropping offshore as a series of reefs including Penguin Island. The 5-kilometre-wide plain in this section contains up to 90 ridges, deposited as the shoreline regressed over the past 6000 years.* ***b)*** *Cross-section of the Lancelin rocky seabed and backing transgressive dune system.*

parabolic dunes on the open coast, particularly between Quinns Rocks and Cape Burny. Generally, smaller transgressive dune barriers occur up to Kalbarri, with the coast from Kalbarri to Steep Point dominated by the Pleistocene calcarenite barrier of the Zuytdorp cliffs, which are in places capped by Holocene, transgressive clifftop dunes.

Within Shark Bay, low-energy, beach-foredune ridges and recurved spits dominate the generally small, low regressive barrier systems. At Carnarvon, the Gascoyne River delivers large volumes of sand to the coast, which on the low-energy, southern side of the river mouth are deposited as a series of beach ridges fronted by mangroves. Extensive sand shoals and spits front the river mouth (see Figure 6.41a) and transport sand northward, where the spits attach to the shore and deliver sand to a 10-kilometre-wide regressive foredune ridge plain that extends to Miaboolya Beach, and contains up to 60 ridges (see Figure 6.41b). As wave energy increases northwards along the beach, these grade into a transgressive barrier with parabolic dunes that have migrated into the southern end of Lake McLeod, with transgressive dune barriers continuing north to Gnaraloo. Once the Ningaloo reefs are reached at Amherst Point, the shelter provided by the reefs leads initially to large regressive barriers, the most extensive located on cuspate forelands formed in lee of reefs. Some of the regressive systems have since been blanketed by transgressive dune activity in the form of long-walled parabolics.

Exmouth to Broome

At Exmouth Gulf the coastline changes considerably. To the north, the coast trends northeast, sheltering it from Southern Ocean swell and resulting in a very

Figure 6.39 *The outer section of the Warnbro foredune ridge plain at Safety Bay.* Photo: A.D. Short

Figure 6.40 *The famous pinnacles in Nambung National Park formed as root casts around tree roots in the coastal dune field – the original surface of the dunes being the top of the casts. When the sand dunes were destabilised, the sand blew away and exposed the casts as the pinnacles.* Photo: A.D. Short

Figure 6.42 *The low-gradient, eastern shore of Exmouth Gulf at Burnside Island. The 1-kilometre-wide fringe of mangroves is backed by the orange-tinged, salt flats that extend 10–12 kilometres to the east.* Photo: A.D. Short

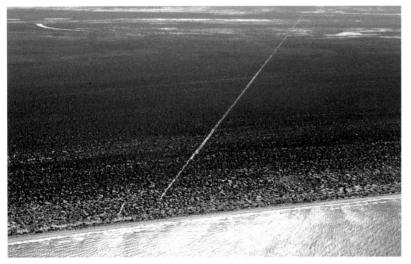

Figure 6.41 *The Gascoyne River delta grades from very low-energy, mangrove-dominated shoreline south of the river mouth to* ***a)*** *the sandy river mouth and low river mouth islands and shoals – north of the river mouth, increasing wave energy has formed the 10-kilometre-wide Miaboolya foredune ridge plain; and* ***b)*** *further north to the parabolic dunes of the Bejaling sandpatch.* Photos: A.D. Short

low-energy shoreline, as the tidal range begins to increase to the northeast. Sediments are supplied by seven rivers draining from the Pilbara. These are activated by periodic tropical cyclones, during which time they can deliver large volumes of sand and mud to the low-gradient coast. The result is a series of low-energy, wave and tide-dominated regressive barrier systems, usually consisting of several low beach to foredune ridges, together with areas of recurved spits, fringed by mangroves and in many places backed by salt flats up to several kilometres wide (see Figure 6.42). This section of coast also contains some of the few barrier islands located on the Australian coast, particularly between Port Weld and Diver Inlet, and in Sherlock Bay.

Evolution of the Pilbara shoreline

Sea level reached its present level along the Pilbara coast about 6500 years ago, when the coast was deeper and less reef-congested and wave energy was higher along the shore. At this time many of the existing beaches, barriers and barrier islands, including those now fringed by mangroves, were deposited. As sediment was delivered to the coast by the Ashburton, Cane, Robe, Fortesque, Yaanyare, Maitland (see Figure 6.43), Harding, Sherlock, Peewah, Yule, Turner and De Grey rivers, the backing lagoons and nearshore gradually shoaled. At the same time, the coral reefs grew to the surface, with both processes lowering waves at the shore and permitting the mangroves to spread. In addition, a slight 1–2-metre fall in sea level further shoaled the nearshore and backing lagoons. Today, the lagoons have largely infilled to form high-tide salt flats, while the shallow shoreline provides extensive intertidal zones for mangrove development that attenuates waves along the shore.

Eighty Mile Beach

Eighty Mile Beach extends for 222 kilometres, between Cape Keraudren and Cape Missessy. It is backed by an equally long barrier system consisting of a southern 70-kilometre-long section from Cape Keraudren to Red Hill, and is characterised by inner and outer discontinuous foredune ridges, some of which are now covered by sand sheets and transverse dunes. The 133-kilometre-long northern section begins north of Red Hill as a series of widely separated, discontinuous foredune ridges and northward migrating recurved spits, backed by salt flats up to 12 kilometres wide. These converge

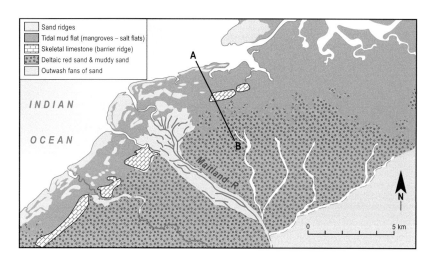

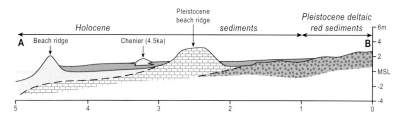

Figure 6.43 *The Maitland River delta is typical of the Pilbara coastal rivers. It consists of a several-kilometre, low-gradient weathered Pleistocene 'inner' delta, including calcarenite Pleistocene beach ridges, and a modern Holocene delta consisting of inner cheniers (dated as formed up to 4500 years ago), intertidal mud flats and an outer, modern low beach-ridge system, which is exposed to the low waves and high tides of the Pilbara coast.*

at the beginning of the final section and form the longest foredune ridge system in Australia, extending unbroken for 73 kilometres to Cape Missessy (see Figure 6.44).

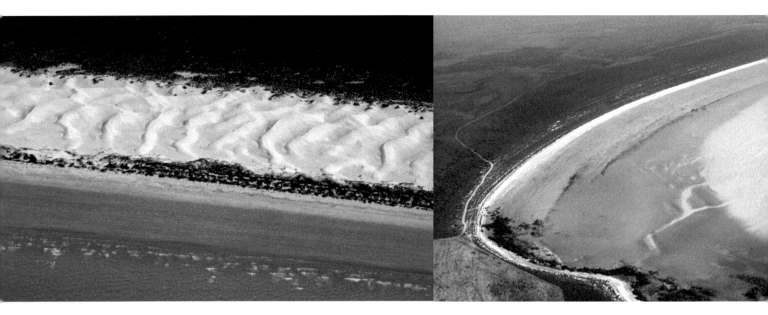

Figure 6.44 *Two views of Eighty Mile Beach:* **a)** *in the south near Cape Keraudren, where it is backed by north-trending transverse dunes; and* **b)** *the very northern end with the ultradissipative sand beach, grading seaward into low-tide mud flats and backed by well-vegetated foredune ridges.* Photos: A.D. Short

Broome to Northern Territory coast

The 9435-kilometre-long northern Australian coast, between Broome and the Northern Territory–Queensland border, contains 31 per cent of the Australian coast, but by area only 5 per cent of Australia's barrier systems. The barriers occupy only 10 per cent of the rugged Kimberley coast, 25 per cent of the western Territory coast and 33 per cent of the Territory's gulf coast, much of the remainder given over to rocky coast and mangroves.

Kimberley coast

The Kimberley coast has 368 barrier systems, which are predominantly short (mean length 1.2 kilometres) and small in area (mean area 0.6 square kilometre). Most are low, low-energy chenier, beach ridge to low foredune ridge systems (see Figure 6.45). There are only four systems larger than 10 square

kilometres, three associated with the few areas of dune transgression between Cable Beach and Cape Leveque in the west, and at Cape St Lambert in the east.

The Northern Territory

The Northern Territory coast has 390 barriers, which occupy 1400 kilometres of coast. Like the Kimberley, most of the barriers are stable to regressive cheniers, beach ridges or foredune ridges, with some of the foredunes experiencing minor transgression. There are several areas of dune transgression, all on exposed, east-facing shores at: De Courcy Head to Brodgen Point and Quoin to Cuthbert points on the north coast, and particularly along the more exposed eastern Arnhem Land coast from Cape Arnhem south to Point Alexander, between Trial Bay and

Point Arrowsmith, and between Muwul-Rantyirrity Point–Wiyakipa Beach.

The east and south coast of Groote Eylandt is also exposed to strong southeast trade winds and waves, and has several extensive transgressive dune systems. The shore is backed by generally stable, northwest-trending, nested-parabolic dunes extending 4 kilometres inland in places. The dunes average 20–30 metres in height, reaching 85 metres in places, some as clifftop dunes.

Figure 6.45 *A typical headland-bound Kimberley beach and low foredune, shown here at low tide, near Freshwater Point, west Kimberley.* Photo: A.D. Short

Conclusion

Australia has an extensive and diverse range of coastal dune and barrier systems, including the world's largest, longest and highest coastal dune systems. The systems, however, range from small, low and low-energy barriers composed of cheniers and beach ridges – particularly along parts of the northern Australian coast where they back sand flats, mud flats and mangroves – to larger systems generally along the exposed sections of high-wave and wind-energy east and southern coasts. Each system is a product of the waves delivering sand to the shore, and with the wind blowing it inland, where in most cases the dunes are subsequently stabilised by vegetation. In total, dune systems back 42 per cent of the Australian coastline, including most of the beach systems. They are therefore a highly prominent and important component of the coastal zone.

While most Australian dunes are covered in dune vegetation and are stable, they are easily de-stabilised and re-activated by vegetation disturbance and removal, particularly along the seaward edge. Management of these systems therefore revolves around maintaining vegetation cover, which in turn supports the associated dune flora and fauna. Some dune systems located near major cities are now covered by residential and tourist developments, and some are mined for sand to build those same developments. The vast majority, however, remain in a natural state, with many located in national parks and reserves. In terms of area, they are our most extensive coastal asset.

ROCKY COASTS

SPECTACULAR ROCKY SHORELINES occupy 40 per cent of the Australian coast. Unlike the more contemporary beach and barrier systems described in chapters 5 and 6, these rocky coasts are the result of ancient rocks that have been exposed to the elements and shaped by marine and atmospheric processes over millions of years. Some of these rocky landscapes have developed into major tourist attractions, such as the Nullarbor Cliffs along the Great Australian Bight in Western Australia, the Twelve Apostles in Victoria and Sydney's North Head, a few of which are presented in this chapter.

What influences how rocks behave in their coastal environment? The morphology of rocks along the coast, rock type, texture, mineralogy, strength, structure, bedding, jointing, permeability, dip, height and slope are all important, and they contribute to the evolution and shape of our rocky coasts. We begin with a brief introduction to the different types of rocks that dominate the Australian coast. After examining the processes operating on rocky coasts and the resulting landforms – particularly the formation of sea cliffs and rock platforms – we offer a conceptual model of rocky coast evolution. The chapter concludes with a tour of some of the more remarkable sections of rocky coast around the Australian margin.

Rock types

In Chapter 1 we summarised the geological evolution of the Australian coast and showed that the rocks composing the modern coast were formed as far back as 3500 million years ago. The most recent period of mountain building and new rock formation occurred 50 million years ago. Today, Australia lies

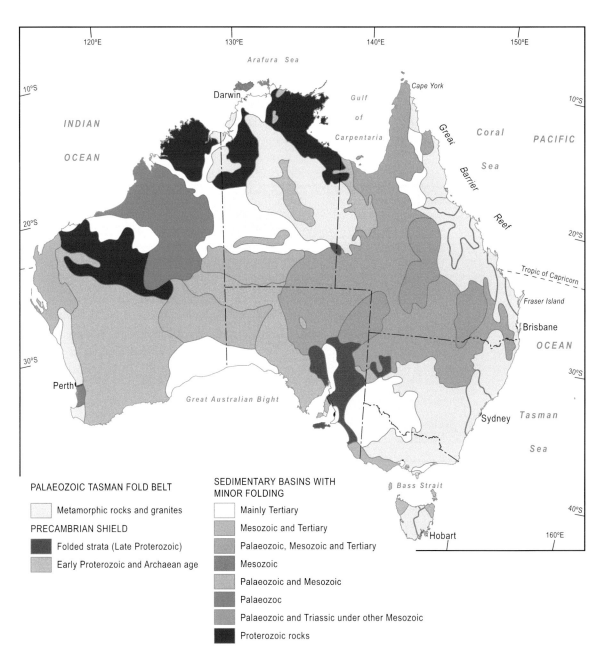

Figure 7.1 Simplified geology map of Australia

PALAEOZOIC TASMAN FOLD BELT

- ☐ Metamorphic rocks and granites

PRECAMBRIAN SHIELD

- ■ Folded strata (Late Proterozoic)
- ☐ Early Proterozoic and Archaean age

SEDIMENTARY BASINS WITH MINOR FOLDING

- ☐ Mainly Tertiary
- ☐ Mesozoic and Tertiary
- ☐ Palaeozoic, Mesozoic and Tertiary
- ☐ Mesozoic
- ☐ Palaeozoic and Mesozoic
- ☐ Palaeozoc
- ☐ Palaeozoic and Triassic under other Mesozoic
- ■ Proterozoic rocks

in a mid-plate setting, with no tectonically active coasts. Nevertheless, rocks still provide outcrops around much of the ancient continent. This is partly due to the recent rise in sea level, which flooded the inner continental shelf and rose to flood valleys and form estuaries, lagoons and beaches, while the intervening hills were eroded to form headlands, as seen, for example, along the Sydney coast. Figure 7.1 is a simplified geological map of Australia; it illustrates the major rock types around the coast, which can be categorised as one of three types: igneous, sedimentary or metamorphic.

Igneous rocks

Igneous rocks derive from molten magma from the earth's interior, and are of two basic types: intrusive and extrusive. Intrusive igneous rocks form under great pressure, deep within the Earth, and cool very slowly. This permits larger crystals to form. The crystals are so tightly packed that there are no pore or air spaces between them. Consequently, they are impermeable. The dominant type is granite, which forms under continents and is composed of sand-sized quartz grains (25 per cent), feldspars (50 per cent) and other minor minerals (25 per cent), including some heavy minerals. The quartz is clear in colour, the feldspars pinkish and the minor minerals that often appear as black specks give granite a mottled, white–pinkish appearance. When exposed at the surface, granite forms massive, rounded outcrops and boulders. As they weather and erode, the feldspars and most minor minerals break down to silt and clay, whereas the quartz and any heavy minerals remain intact as highly resilient sand grains.

Extrusive igneous rocks form at or near the Earth's surface, usually as a result of volcanoes

Figure 7.2 *The dolerite-dominated Cape Bruny, on Bruny Island, Tasmania.*
Photo: A.D. Short

erupting magma, which flows as lava and cools as fine-grained basalt or medium-grained dolerite. Because it is exposed at the surface, basalt cools rapidly and forms very small crystals. Basalt cooling just beneath the surface may contract as it cools, forming distinctive vertical columns. It is filled with air bubbles and is highly porous, or permeable. Basalt is composed of dark minerals such as pyroxene and olivine, both of which weather rapidly to form silt and clay, but no sand-sized crystals other than rock fragments. Dolerite is very similar but is intruded just below the surface and consequently cools more slowly. As a result, dolerite is coarser grained, highly resilient and can also cool to form large columns (see Figure 7.2). Igneous rocks are the source from which most other rock types are derived, with their eroded debris contributing to the second major type: the sedimentary rocks.

Sedimentary rocks

Sedimentary rocks are rocks that contain material that has been eroded, transported and deposited as sediments. The sediments were usually transported by rivers or glaciers and deposited in estuaries, deltas, the continental shelf or deep ocean. As they are buried under subsequent layers of sediment, the weight, pressure and heat slowly cement and transform the unconsolidated sediment into consolidated rock, which is classified by the size of the sediments into: mudstone, siltstone, sandstone and conglomerates (a mixture of coarser grains). They are also classified by their dominant mineralogy, such as limestone (carbonate-rich), coal (humic-rich) and pelagic (fine-microscopic organisms). Sedimentary rock may also contain fossils buried along with the sediment.

Sedimentary rocks are laid down in layers or beds of sediment, ranging from a few millimetres to tens of metres thick. They frequently have vertical cracks, called 'joints', which commonly trend in a couple of preferred directions. Furthermore, sedimentary rocks are often porous, with water penetrating either within the rock fabric, especially in unfilled pore spaces between the grains, or along bedding and joint planes, both of which can form lines of weakness. Their horizontal beds, vertical joints and porosity are extremely important in determining how sedimentary rocks respond to coastal processes.

Metamorphic rocks

The third major rock type is metamorphic, which includes any rock type that once was altered by high heat and pressure, or 'metamorphosed'. If the former rocks were sedimentary, then they transform into metasedimentary rocks. Metamorphism typically occurs:

- when the rocks are buried under deep layers of sediment (burial metamorphism)

- where they are compressed by plate movement (regional metamorphism)

- in proximity to molten igneous rocks (contact metamorphism).

The result is a partial recrystallisation and segregation of the minerals that formed the rocks. Metasedimentary rocks include fine-grained rocks (mudstones and siltstones), which become slate; coarser sandstones become gneiss, while pure quartz sandstone becomes quartzite and limestone transforms into dolomite and then marble. As a result of metamorphism, the rocks become less permeable and ultimately impermeable, like the granites. In addition, compression metamorphism, particularly in fold belts, can lead to a buckling of the originally horizontal sedimentary rocks, so they can be tilted and twisted and potentially turned 180 degrees, with the rock dipping at angles ranging from horizontal to vertical.

Australian rock types

All of the above major rock types and other secondary types occur around the Australian coast. The type of rock at any particular location depends on the geological history of the area. In this section we briefly outline where the different rock types occur, in each case beginning in the northeast and working clockwise around the coast.

Granite

Granite headlands and rocky sections are scattered down the east coast, forming several high islands

Figure 7.3 *Massive sloping granite at Remarkable Rocks, Kangaroo Island, in South Australia. The cars parked in the car park provide an idea of the scale of these rocks.* Photo: A.D. Short

in Torres Strait and many of the prominent headlands along northeast Queensland between Temple Bay and Bowen, including Cape Melville. Lizard Island, Magnetic Island and many of the Whitsunday islands are granitic. Granite outcrops at numerous other sites along the southeast coast, such as around Moruya (where it was quarried to create the Sydney Harbour Bridge pylons), Tuross Heads and Bega in New South Wales; Gabo Island and Wilson's Promontory in Victoria; and several of the Bass Strait islands, including much of Flinders and Cape Barren islands. It also dominates the northeast Tasmanian coast between Anderson Bay and Freycinet Peninsula–Schouten Island, including the boulder-strewn coast of the Bay of Fires, and sections of the west coast.

In South Australia, granite outcrops on the southwest coast of Kangaroo Island, forming the famous Remarkable Rocks (see Figure 7.3), as well as Cape Willoughby and Cape du Couedic at the eastern and western ends, respectively. It dominates the ancient Gawler Block that forms the basement of most of the western Eyre Peninsula, outcropping as prominent headlands between Cape Catastrophe and Cape Adieu.

Granite is at its most extensive along the southwest of Western Australia, where the ancient Yilgarn Block dominates the coast between Point

Dempster and Esperance, and between Albany and Cape Naturaliste. In the northwest it also dominates the ancient Pilbara block, forming much of the rocky coast between Cape Preston and Cape Thouin. Granite is absent from the basin-dominated northern Australian coast, apart from a small area on the eastern Arnhem Land coast in Melville Bay and between Point Alexander and Cape Grey.

Basalt

Basalt is usually associated with past volcanic activity and tends to be more dispersed along the east coast, formed as the continent moved northward over a volcanic hot spot. Most basalt in Australia is located along the Great Dividing Range, with only a few outcrops at the coast. Most notably, it appears on the coast in southeast Queensland, where it forms the border at Point Danger extending into northern New South Wales at Fingal Head, Broken Head–Lennox Head and on the southern New South Wales coast between Kiama and Gerroa, and at Broulee–Congo. A similar hot spot beneath oceanic crust in the Tasman Sea formed the basaltic Lord Howe Island. Norfolk Island, further to the northeast, was also formed from horizontal basalt flows.

In Victoria, basalt forms Australia's youngest rocks at Killarney Beach, where around 25 000 years ago the Tower Hill volcanic eruptions resulted in molten lava flowing several kilometres to the coast and forming a series of low basalt outcrops between The Basin and Reef Point (see box 'Australia's most recent volcano', page 8). Basalt and dolerite are most extensive in Tasmania, where Jurassic dolerite sheets covered most of the eastern half of Tasmania. On the north coast of Tasmania, it outcrops between Stanley and Low Head, including well-known sites such as The Nut and Table Cape, both former volcanic plugs. Basalt dominates the east and southeast coast between Swansea and South Cape, forming the spectacular Cape Pillar – the 'pillars' are actually columnar dolerite. Although basaltic rocks are absent in South Australia and most of Western Australia, basaltic lavas flooded the Dampier Peninsula and the Kimberley region 1800 million years ago. Today, it outcrops at the coast around Wilson Point, Port Nelson, in Admiralty Gulf and most prominently around Cape Londonderry, the northernmost tip of Western Australia.

Sedimentary rocks

Sedimentary rocks are the most prolific rock type around Australia, occupying all the sedimentary basins (see Figure 1.5). They tend to be horizontally bedded, deeply jointed and range from weaker shales to more resistant sandstones. There are some areas of more erodable and soluble limestone and marls, and the younger Pleistocene dunerock, known as 'calcarenite' (see Chapter 6).

Along the Queensland coast, sedimentary rocks occupy the northern tip of Australia at Cape York and the Laura Basin, centred on Princess Charlotte Bay; both are dominated by mudstone, siltstone and sandstone. The Bowen Basin at Townsville and Proserpine Basin in Repulse Bay are dominated by silt to sandstones; Styx Basin at Broad Sound by silt to sandstones; and Maryborough Basin, centred on Hervey Bay, by sandstone and conglomerate.

In New South Wales, sedimentary rocks outcrop in the Clarence–Moreton Basin (silt to sandstone), between the Gold Coast and Evans

Head. They dominate the sandstone-rich Sydney Basin between Newcastle and Durras (see Figure 7.4), which is underlain by sedimentary coal deposits, outcropping at the coast around Newcastle and at Coalcliff, north of Wollongong.

In Victoria, sedimentary rocks form the weaker Tertiary mud to sandstones and some limestone of the Gippsland and Otway basins. The Gippsland Basin dominates the eastern Victorian coast, whereas the Otway Basin occupies the western coast and includes all the rock types visible along the Great Ocean Road, including more resilient sandstones to the east of Cape Otway and the more rapidly eroding, horizontally bedded marine marls that form the Twelve Apostles to the west of the cape.

Sedimentary rocks are limited to the northwestern corner of Tasmania, between the Pieman River and Boat Harbour, where ancient sandstones and mudstones form the coast at Rocky Cape. There are also a few limited exposures along the central north coast near Burnie, Ulverstone and Badger Head, and on the west coast just south of Macquarie Harbour.

In South Australia, the Murray Basin has filled with dolomite in the southeast, between the Victorian border and Cape Banks, and Quaternary marine and aeolian sands, which are partially lithified into calcarenite between Cape Banks and Middleton (see Figure 5.33). Smaller sediment-filled basins occupy St Vincent and Spencer gulfs, with sandstones forming the low coasts in the upper regions of both gulfs. The most dramatic sedimentary rocks are the horizontally bedded limestones, exposed as the spectacular, 90-metre-high Bunda Cliffs, which extend for 210 kilometres along the uplifted Nullarbor Basin.

The Western Australian coast is dominated by five large basins, all filled with sedimentary material. In the southeast is the western section of the Nullarbor Basin, between Eucla and Israelite Bay, and whose 90-metre-high limestone cliffs form the 160-kilometre-long Baxter Cliffs. The Perth Basin extends from Augusta north to Geraldton and is filled with sandstone, siltstone and limestone. Immediately to the north is the Carnarvon Basin, between Geraldton and Cape Preston, most renowned for the red sandstone cliffs that occupy the Murchison Valley and form the Kalbarri coast. It also includes the 300-kilometre-long Zuytdorp Cliffs (see box 'Zuytdorp Cliffs', below and Figure 7.5), up to 150 metres high, which are composed of lithified sand dunes (calcarenite) and trend north to form Dirk Hartog, Bernier and Dorre islands. The

Figure 7.4 *Horizontally bedded sedimentary rocks at Macquarie Light, Sydney (sandstone).*
Photo: A.D. Short

ZUYTDORP CLIFFS

The Murchison River mouth at Kalbarri marks a major change in the geology, orientation and nature of the west coast. The Silurian red sandstones that dominate the hinterland and coast to the south as far as Bluff Point are replaced by the most massive, continuous Pleistocene dune calcarenite deposits in the world. It commences at the river mouth and extends to the northwest initially as steep 100 to 200-metre-high cliffs, for 176 kilometres as Zuytdorp Point, part of the way to the Zuytdorp Cliffs. The cliffs then turn and trend more northerly for another 34 kilometres to Steep Point, the westernmost tip of the Australian continent, before breaking into a series of calcarenite islands that continue north for another 155 kilometres, and finally terminating at Cape Ronsard on Bernier Island. In all, the deposits extend for 365 kilometres. The cliffs represent successive layers of carbonate-rich, transgressive dune sands that have accumulated over the past 2 million years during each sea-level highstand. During periods of lower sea level, the dunes are partially lithified (cemented) into calcarenite (see Figures 6.16 and 7.5).

uplifted and deeply weathered rocks of Exmouth Peninsula's Rough Range are composed of sandstones overlain by limestone. The low-gradient, sedimentary shore between eastern Exmouth Gulf and Cape Preston continues to be supplied with unconsolidated sediments from the Pilbara river systems. The Canning Basin between Port Hedland and King Sound is dominated by sandstones to Cape Keraudren, then Eighty Mile Beach, and then the generally low, deeply weathered sandstone bluffs of the Dampier Peninsula that commence at Broome. Finally, the ancient Kimberley Basin is largely composed of bright-red, horizontally bedded and deeply weathered and dissected sandstones.

The Joseph Bonaparte Basin straddles the Western Australian and Northern Territory coast

Figure 7.5 *Active, long-walled parabolic dunes on Dirk Hartog Island.*
Photo: A.D. Short

between Cambridge Gulf and Fog Bay. However, only a few low sandstone bluffs outcrop at Western Australia's Cape Domett and between Fossil Head and Jenny Point in the Northern Territory. Tertiary sedimentary rocks of the Monkey Shoals Basin form the generally low northern coast around Darwin, between Cox Peninsula and Cape Hotham, as well as Bathurst and Melville Islands and the Coburg Peninsula, where they generally outcrop as low, lateritised bluffs. Laterites – formed as humid tropical soils, with accumulation of oxides of iron and aluminium – have an upper reddish crust underlain by a white, leached and pallid zone. In Kakadu National Park, Kombolgie sandstone of the Pine Creek Orogen forms the spectacular escarpments. The Arnhem Land coast is dominated by the sedimentary rocks of the Arafura Basin in the north, and McArthur Basin from Blue Mud Bay down to the Queensland border. Where exposed at the coast, both basins provide usually low, lateritised sedimentary sandstone, siltstone and some conglomerates.

The Gulf of Carpentaria and the Carpentaria Basin dominate all of the Queensland Gulf coast, where – when exposed at the coast – outcrops are low bluffs of deeply lateritised soil, formed on a mixture of sedimentary rocks.

The youngest sedimentary rocks on the Australian coast are also the most widespread. They consist of partially lithified, carbonate-rich Quaternary coastal sand dunes – the rock known as 'aeolian calcarenite' or 'aeolianite', and more commonly called 'dunerock' (see Figure 7.6). Calcarenite is widespread along the southern and western Australian coast, and is discussed in Chapter 5.

Figure 7.6 *Calcarenite cliffs at Whalers Way, South Australia, are composed of multiple layers of lithified dune sands and soils.* Photo: A.D. Short

Metamorphic rocks

Metamorphic rocks are the second-most common rock type in Australia. They are associated with the orogenic zones, where sedimentary and other rocks have been buried, buckled and deformed during a phase of mountain building. They therefore occupy sections of coast where fold belts dominate. The massive Tasman Fold Belt (see Figure 1.4) dominates the east coast between Cape Melville and Kangaroo Island. The belt consists of a 1000-kilometre-wide series of ancient, and now heavily denuded, fold belts. It contains folded, more resilient metamorphic rocks that are separated by the usually less-resilient and horizontally bedded sedimentary basins described above.

On the east coast, metamorphic rocks commence in the Hodginkson–Broken River fold belt, which occupies the coast south of Cape Melville to Hinchinbrook Island, while the massive New England Fold belt continues south from Cape

***Figure* 7.7** *Steeply dipping, metasedimentary rocks truncated by the rock platform at Hallett Cove, near Adelaide in South Australia.* Photo: A.D. Short

Upstart for 1600 kilometres to Port Stephens in New South Wales. The Lachlan Fold Belt commences at Durras on the New South Wales south coast, dominates the coast down to the border at Cape Howe, and continues along the eastern Victorian coast to Lake Tyres. It also outcrops along parts of the northeast Tasmanian coast, between Low Head and Bicheno. Many of the outcrops are composed of steeply dipping slates and gneiss.

In South Australia, metamorphic rocks are associated with the Adelaide Geosyncline or Fold Belt, which forms most of the coast of Gulf St Vincent, part of northern Spencer Gulf and northern Kangaroo Island. The dipping folded rocks are best displayed where they are truncated by rock platforms, as at Hallett Cove (see Figure 7.7).

In Western Australia, metamorphic rocks

are restricted to the Cape Lambert region of the Pilbara coast. A second, smaller section of folded rocks associated with the King Leopold Fold Belt, which borders the southern edge of the Kimberley basin, forms the highly indented coast and scores of elongate islands between King Sound and Collier Bay, including the Buccaneer Archipelago.

In the Northern Territory, the only metamorphic rocks are associated with the Pine Creek Orogeny, which extends to the coast either side of Darwin, and is also exposed in parts of Kakadu National Park.

Rocky coast processes and morphology

In the preceding three chapters we have shown that landforms are shaped by the dominant processes affecting them: estuarine sediments are shaped by the relative influences of wave, tide and river processes; beaches are controlled by waves and tides that mould the soft, mobile sand; and dunes require strong winds to form the largest systems. Rocky coasts, although exposed to the same wind, wave and tidal processes, react much more slowly. While the eroding force of waves is the most obvious, there is a range of additional factors that play a role in the formation of rocky coasts. Among these are the nature of the rock and the geochemical interaction of marine processes with the rock fabric, as well as secondary factors such as tidal range, coastal orientation and inheritance from former shorelines.

Rocky coasts can adopt a wide range of shapes and profiles. They can range from vertical plunging cliffs, to cliffs fronted by near-horizontal rock platforms, across a range of slopes in between. It is best to consider rocky coasts in terms of three zones (see Figure 7.8): the spray zone above the reach of direct wave attack – usually consisting of a bare or sparsely vegetated cliff or slope; the breaking waves zone, which is directly impacted by waves and swash processes extending from just below sea level to a few metres above; and the subtidal zone below low water. Each of these zones is exposed to a different range of physical, chemical and biological processes, which together with the nature of the rock determine the morphology of a particular section of rocky coast. For simplicity, we discuss them here as the subaerial, intertidal and subtidal zones, although their boundaries need not coincide with these tidally defined zones.

Zones of erosion

The subaerial zone

The subaerial zone extends from the upper limit of regular wave attack to the limit that is regularly affected by salt spray. Depending on the level of wave and wind energy, this zone tends to be a few to tens of metres above sea level. It is exposed to the same atmospheric processes that dominate the hinterland. However, there are two important additional factors:

Figure 7.8 Waves break over the platform and reach the base of the notched cliff at Zuytdorp Point, Western Australia. During such events, debris is removed from the base of the cliff. Note the wave-deposited boulder (arrow) atop the 30-metre-high cliff – a common sight along this exposed, high-energy coast. The cliff dominates the subaerial zone and the wave-washed platform the intertidal zone, while broken waves and white water cover the subtidal zone. Photo: A.D. Short

the input of salt spray that regularly drenches the zone, and thereby accelerates corrosion, and the usually over-steepened coastal slopes, which – on longer time scales – can form steep cliffs where gravity delivers debris to the base of the slope. Both have a dramatic impact on the erosion and evolution of the rock face.

Combined, these processes have three major impacts. First, the normal subaerial processes of precipitation, temperature variations and wind act on the surface. These are aided by the salt spray, which adds a rich cocktail of salts, particularly sodium chloride. The salt and wind modify and denude the vegetation, while the spray saturates the soil with salt, which accelerates the oxidation of unstable minerals in the rock fabric – the equivalent of 'rusting' the rock. The result is an accelerated geochemical

disintegration of the rock's fabric and strength. The surface usually has a sparse covering of salt and wind-tolerant vegetation, all of which expose the soil and rock to more direct attack and permit it to weather and erode more rapidly. At this stage, the second impact of salt water comes into play. The continuous drenching by sea spray lubricates the rock surface with water, which assists its movement down-slope. Finally, the weakened soil and/or rock, under the force of gravity and assisted by the lubrication of sea spray, falls, slides, slumps or creeps down-slope to the intertidal zone. Over time, this accelerated weathering leads to erosion and an over-steepening of the slope, thereby further accelerating the gravitational processes. In addition, as material is eroded and moved down-slope, it exposes fresh rock face and minerals to the salt spray and the weathering enables the erosion to continue at an accelerated rate.

Together, these processes accelerate the erosion and oversteepening of the seaward slope, the final angle of the slope depending on the processes, plus the nature of the rock. Simply put, the more resistant and massive the rock (granite, basalt, metamorphic, sandstone), the steeper the slope. Weaker rocks (shales, mudstone, moraine) tend to slump, providing a gentler slope. This is why rocky slopes can range from cliffs to more gently sloping terrain.

The intertidal zone

The second zone, the intertidal zone, is the most energetic of the three. In this zone, all the above processes, plus breaking waves, tides, currents and biological activity, impact the rock surface. The intertidal zone extends from low tide, up to a few metres above sea level on higher-energy coasts.

Breaking waves provide physical energy to do work, to erode and move rocks – even large rocks weighing several tonnes – in a process called 'wave quarrying'. In extreme conditions, waves breaking on rocks and crevasses exert tremendous hydraulic pressure, leading to massive rock failure and the erosion or movement of large slabs of rock. The breaking waves and their currents are also the major mechanism for moving debris from the base of the slope and transporting it longshore to build boulder beaches, or seaward to form piles of rock debris in deeper water off the edge of the cliff or platform. While waves do remove and transport the debris, it is the accelerated cliff erosion, described above, that is the prime cause of the erosion, not the waves. The waves and associated currents also carry smaller rocks, called 'toolstones', which gradually abrade the rock, performing a superficial polishing of the rock surface. This contributes little to the overall erosion, but hastens their own attrition. The intertidal zone is also permanently saturated by water, while the rocks in the subaerial zone immediately above are alternately wet and dry. The boundary between the intertidal and subaerial zone is called the 'level of saturation'; it defines the upper limit of the 'intertidal' zone, although this level may be a few metres above sea level in exposed locations, as is described below.

The intertidal zone is also dependent on the tidal range, which determines the position of the shoreline. In increasingly higher tide ranges, the intertidal zone becomes more extensive. Both the waves and tides generate currents, which can move across the rock surface and move small toolstones to abrade the surface and from which debris is gradually removed. Finally, the intertidal zone hosts a rich ecosystem, described in Chapter 3.

Some organisms may protect the rocks by covering their surface and maintaining moisture, while others, such as sea urchins, actively graze and erode the rock surface.

The subtidal zone

The subtidal zone extends from low water to a depth at which wave and tidal processes are unable to erode the rock surface. This depth increases with increasing wave energy, and ranges from less than 1 metre on low-energy shores to as much as 20–30 metres on high-energy coasts. This is a zone in which the physical impact of the waves is responsible for most erosion, and is assisted by gravity, along with some surface erosion by grazing organisms.

Role of lithology

Lithology refers to the physical and chemical character of a rock. The massive, intrusive granites and most of the metamorphic rocks are highly resilient to erosion because they have no pore spaces and are impermeable, so erosion is restricted to the rock face. In coastal granite, this leads to exposed faces that have been stripped of their soil, but maintain their original shape and slope directly into the sea, with no evidence of additional marine attack (see Figure 7.3).

The metasedimentary rocks, while impermeable, commonly have joint and bedding planes inherited from the former sedimentary rocks. These induce lines of weakness that can be etched out by marine attack. Metasedimentary rocks are commonly distorted and may dip vertically into the sea. In the presence of marine attack, they tend to form very jagged shorelines (see Figure 7.7).

Figure 7.9 *The 90-metre-high Nullarbor cliffs are composed of white, friable Wilson Bluff limestone overlaid by the orange–grey Nullarbor limestone, with a narrow rock platform at the base. The vehicle on the cliff top provides an indication of scale.* Photo: A.D. Short

The extrusive igneous basalt and all sedimentary rocks are porous and soak up water and seawater. This has two major impacts at the coast. First, within the rock there is a water table, below which the rock is always saturated, and the rock remains wet with no free oxygen to permit oxidation of the rock fabric. Here, the rock can only be eroded by physical attack of waves, currents and abrasion. Above the level of saturation, it is alternately wet and dry, with a rich weathering mixture of minerals, water, oxygen and salt, which leads to accelerated geochemical disintegration of the rock fabric. Basically, it oxidises more rapidly – more commonly known in some minerals as rust.

Sedimentary rocks often also contain distinctive horizontal and vertical structures, which influence cliff morphology. Horizontally bedded units may erode differentially, with protruding harder and eroded weaker layers, while joints and faults enable water to seep deeper into the rock, and usually determine where breakages occur (see Figure 7.9). These two lines of horizontal and vertical weakness predetermine the zones of erosion and the size of the eroded slabs of rock, which can range from small pieces of shale to large blocks of sandstone.

Rock cliffs, platforms and terraces

The three zones of rocky coast erosion can, in turn, produce three distinct rock surfaces. The subaerial zone is usually dominated by cliffs or over-steepened slopes; in the intertidal zone there may be a near-horizontal or gently sloping rock platform, usually located above sea level; and in the subtidal zone the base of the cliff, platform or slope marks the base of wave erosion, sometimes forming a terrace.

Sea cliffs

Sea cliffs and steeper slopes tend to form in two environments controlled by lithology. In resistant, homogenous granites and resilient metamorphic rocks, marine erosion removes the soil and covering rock but does not erode or over-steepen the slope. The slope, therefore, is dependent on the pre-existing surface, which may be steep or gentle. When steep, the erosion forms cliffs, with the

vertical face maintained by rock falls. In sedimentary rocks, basalts and some weaker metamorphic rocks, marine erosion wears away the base of the subaerial rock surface and over time over-steepens the slope. In this case, the cliffs are a direct result of marine attack. Most Australian cliff systems form in:

- sedimentary rocks, as along the Sydney coast (sandstone/shale), the Port Campbell coast (marine marls) in Victoria, the Nullarbor in South Australia and Baxter cliffs (limestone), Zuytdorp Cliffs (calcarenite) and the Kimberley coast (sandstone-conglomerates), all in Western Australia
- metasedimentary rocks, such as along the Bermagui coast in New South Wales
- basalt, as along the Kiama coast in New South Wales, and on Lord Howe Island.

Some steep slopes and cliffs do occur along the scattered sections of granite coast, with the most extensive occurring along the south coast of Western Australia, particularly in the Esperance and Albany regions.

Rock platforms

On some rocky shores, a near-horizontal platform is located at the base of the cliff or slope. These are most appropriately termed 'rock platforms'. In some places, particularly where there is a substantial tidal range, the platform may be more a sloping ramp than a platform. Elsewhere, the platform is surprisingly horizontal and may even have a raised rampart at its outer margin, and slope landward. Rock platforms are sometimes called 'shore platforms', and have been described as 'wave-cut platforms', but as we explain below, waves are unlikely to be the sole agent in eroding the platforms. Rock platforms are usually backed by a cliff and fronted by a drop-off into deeper water, and many contain a number of secondary features including cliff debris, notches, channels, cuesta (dipping rocks), potholes and ramparts (see Figure 7.10).

Water-layer levelling

As we discussed above, the intertidal zone is where permeable rocks tend to be saturated – the actual level of saturation being dependent on sea level, wave height and rock permeability. Saturation is likely to

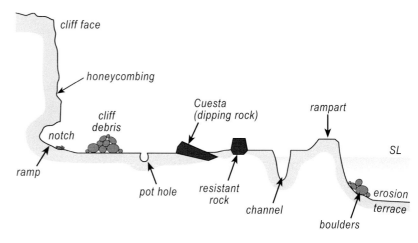

Figure 7.10 An idealised rock platform with its roughly horizontal surface rising to a seaward rampart, and also containing a variety of secondary features produced by erosion, solution and rock type.

Figure 7.11 At Mosquito Bay in New South Wales, intertidal rock platforms are cut into steeply dipping metasedimentary rocks. Photo: A.D. Short

be lowest in areas of low-wave energy and highly porous rocks, reaching as low as mean sea level, and higher in more exposed locations where waves surge up the rocks and elevate the level of saturation by up to several metres, particularly in less permeable rocks. As a consequence, the intertidal zone can range from around mean sea level in some porous limestone, to as much as several metres above sea level in some more resistant sandstones, basalts and metamorphic rocks (see Figure 7.11). This level determines the elevation above which the overlying rocks are exposed to alternating wetting and drying, and the resulting accelerated action by water, oxygen, minerals and salt, which transform the rock crystals to weaker, flaky oxidised forms that can crumble or be removed by gravity and/or washed away by waves. The process of accelerated rock erosion above the level of saturation on the surface of rock platforms is called 'water-layer levelling'; the overlying rock is weathered or eroded down to the water level, or level of saturation.

Salt crystallisation

A second process occurring above this level is salt crystallisation, a process whereby water evaporates, leaving salt crystals. These crystals can force grains of rock apart, thereby gradually weakening their fabric. As the overlying rocks are eroded and their weight is removed, the release of pressure leads to rock slackening, a natural expansion in the rock fabric. This further permits marine attack along the expansion joints and beds. The end result of all these processes is that some minerals are dissolved and others oxidised, and the rock's fabric is chemically and physically weakened. With the aid of gravity, the weakened rock falls into the sea or to the base of the cliff, exposing more fresh rock to attack. The debris at the base of the cliff is then removed by waves at high tide and/or by storm waves and associated currents. The waves remove the debris, but do little if any direct erosion of the cliff, particularly when it is fronted by a platform tens to hundreds of metres wide.

Ramparts and platform elevation

Two other features occur on rock platforms that are a product of erosion along the level of saturation. The first is called a 'rampart', which is an elevated section of the platform usually along its seaward and most exposed margin, behind which the backing platform slopes landward. The reason for this is simple: the outer edge of the platform is exposed to the highest waves and swash attack and, consequently, is constantly wet and has a super-elevated level of saturation. The rear of the platform, however, which may be only occasionally reached by waves, has a lower level of saturation. Therefore, erosion of the cliffs begins at the outer high level, and as it retreats the erosion surface slopes landward toward the rear lower level. This is further evidence that physical wave erosion is ineffective, since the highest part of the platform is located where waves are highest and the lowest is well away from wave attack (see Figure 7.12).

Second, the elevation of the platform varies as a result of exposure to wave activity. Around an embayment, the level of saturation will be highest on the outer exposed part of a headland, decreasing as wave height decreases into the more protected, adjoining bay. As this occurs, the base of cliff erosion and the level of the platform also decreases in elevation, from areas of high to lower waves, a

Figure 7.12 *A rampart at Mosquito Bay (steeply dipping metasediments). Note how the highest part of the platform is located at the exposed seaward edge.* Photo: A.D. Short

feature commonly observed when you walk from a sheltered bay shoreline onto an exposed headland.

Secondary processes

Once the platform begins to form as a roughly horizontal to slightly landward-dipping surface, a number of secondary processes act on the surface and along joints and bedding. The first is physical abrasion by toolstones. These gradually abrade the stones, smoothing off the rough edges and polishing the rock surface, and in some locations wearing hollows (potholes) into the rocks.

The second process is chemical. Any undulations across the rock surface produce secondary, or minor, variations in surface elevation, so that the higher parts of the rock dry faster than lower, more saturated levels. This leads to secondary water-layer levelling. Some intricate surface bumps and hollows can be formed in this process (see Figure 7.13). Similarly, on the cliff face, chemical etching often leads to homeycomb weathering, whereby chemical solution of the rock forms a series of depressions, like the dimples on a honeycomb. On limestone coasts, the solution is an important process, leading to an etching of the rock surface in the upper intertidal zone that produces fretted surfaces called 'lapies'.

Finally, biological activity plays an important, although not always apparent, role on rocky shores. Organisms can be both agents of erosion and protectors. They can even accrete material onto the rock. The benthic organisms that live on the rock surfaces cover the rock and reduce evaporation, keeping the rock saturated. They also increase surface roughness, thereby reducing current strength, and directly protect the rocks from wave attack and

Figure 7.13 *Solution pits on the sandstone platform at Avoca, in New South Wales.* Photo: A.D. Short

abrasion. Other mobile organisms, including grazers such as chitons and gastropods (snails), move over the rock surface to feed on microalgae and, in doing so, can abrade the rock surface with their radula, a rasp-like structure used for feeding. The most intrusive are the boring organisms, such as some molluscs and sea urchins, which bore hollows into rock so that their delicate shells are protected against predators.

Wave erosion

Waves play a very important role in rocky coast erosion. Wave spray determines the upper level of the salt-drenched, subaerial zone and, along with sea level and tides, the upper level of saturation of the intertidal zone. Waves also are the major physical process in the intertidal and subtidal zones. In the intertidal zone, while waves may take a secondary role to geochemical erosion of the cliff face where platforms are present, they are required to wash away and remove all the debris delivered by rock falls and slides from the cliff, and the products of geochemical processes. Waves transport large debris as bedload, resulting in the movement of toolstones and the abrasion of the rock surface.

Most importantly in the lower intertidal and subtidal zones, where wave breaking and energy is highest, waves do physically attack the seaward edge of the cliff or platform. In sedimentary rock with bedding and jointing planes, in rocks of differing strength, and in columnar basalt, the hydraulic force exerted by the breaking waves gradually erodes along the bedding and joint lines to quarry out a piece or slab of rock, a process called 'wave quarrying'. The size of material eroded by wave quarrying can range from small fragments of shale to large sandstone boulders weighing several tonnes. The wave quarrying erodes the seaward face of the cliff or platform, to a depth of major wave agitation as shallow as 1 metre on lower-energy sections down to a depth of 20–30 metres on high-energy coasts. This lower level of wave truncation is the true base of the cliff and elevation of the subtidal erosion terrace.

Rock platforms form only in suitable rock types – usually sedimentary, basalts and some meta-sedimentary rocks – where the rate of geochemically accelerated cliff retreat is greater than the rate at which waves quarry, abrade and erode the outer edge of the platform. If the rate of cliff retreat exceeds that of the quarrying, the platform will form and widen. Since wave quarrying tends to be greatest on the most-exposed section of a headland, platforms tend to be too narrow here, widening in more moderate energy locations. At the same time, the exposed platforms will be higher, due to the raised level of saturation, and lower in more moderate and sheltered locations.

Very few measurements of rock platform dimensions exist. However, regional studies in South Australia show that along the 1600-kilometre-long coast of Eyre Peninsula rock platforms cut primarily in calcarenite average 40 metres in width, with a very few extending more than 100 metres seaward. Similar measurements on Kangaroo Island found platforms cut into calcarenite averaged 64 metres in width, with a maximum width of 120 metres; while those cut into more ancient and resilient metasedimentary rocks averaged between 30 and 40 metres in width. Based on these figures, the mean rate of cliff retreat required to expose the platforms ranges from 5 millimetres per year for the older and more resistant metasedimentary rocks to 10 millimetres per year in the softer calcarenite. Similarly, on the Sydney coast platforms cut into the massive Hawkesbury sandstone average 40 metres in width and usually lie a few metres above sea level. In contrast, those cut in the softer, more porous Wianamatta shales lie between low and high water and average 120 metres in width, which would suggest maximum rates of cliff retreat in the order of 6 millimetres and 20 millimetres per year, respectively.

Arches, sea stacks and blowholes

Several distinctive and often-attractive landforms have developed in the erosion of rocky coasts, including arches and sea stacks, blowholes and old

Figure 7.14 Natural
Bridge, a sea arch cut
in massive granite,
near Albany in Western
Australia. Photo: A.D.
Short

hats. Sea arches can form when a more resistant rock overlies a less resistant rock type; the erosion of the less-resistant rock leaves the overlying rock as an arch. This is the case at Admiralty Arch on Kangaroo Island, where weaker calcarenite overlies resistant granite; and an arch through Roach Island off Lord Howe Island is excavated along a dyke. Arches can also form in heavily jointed massive rocks when lower blocks are quarried out, leaving the overlying rock, as occurs at Natural Bridge near Albany in Western Australia (see Figure 7.14).

Sea stacks tend to form in deeply jointed rocks, particularly when there is a line or zone of weakness in some of the joints. This enables wave quarrying to remove the weaker zones, leaving behind the more resistant core as a stack. The Twelve Apostles in Victoria are Australia's best-known stacks. A double arch characterised London Bridge on the Great Ocean Road, until collapse left the outer part as a sea stack with only the single arch.

OLD HAT PLATFORMS

Old hat platforms refer to a small island with a rock platform surrounding the island, giving the appearance, particularly at low tide, of an old battered hat with a brim. Such platforms occur to the north and south of Windang Island off Wollongong, and at Soldiers Cap, an old hat island 150 metres off the east coast of Lord Howe Island. In general, old hat platforms are highest and narrowest on their more exposed side, and lower and wider on their more moderate and lower-energy sides.

Blowholes tend to form in rocks with zones of internal weakness. The famous Kiama Blowhole in New South Wales is related to a lava tunnel left in the basalt. The Quobba blowholes in Western Australia have been hollowed out in the weaker and soluble calcarenite.

These unusual features of rocky coasts tend to be the aberrations, produced by an unusual combination of factors, particularly differential erosion leading to the arches, stacks and caves. They are, however, those features which tend to attract public interest on the scenic coasts along which they occur, and in cases where there is reasonable access, become local and even international tourist attractions, such as the Kiama and Quobba blowholes, the Twelve Apostles (sea stacks), the Tessellated Pavement (wave quarrying), Nullarbor Cliffs (sea cliffs) and Natural Bridge (sea arch).

Inheritance from past sea levels

When examining the present-day rocky coasts, consider that most of the Australian coast was exposed to similar erosion processes during past highstands of the sea, at or close to present level, the most recent 120 000 years ago being 2–3 metres above present sea level around much of the coast. Some of the erosion of rocky landforms seen at the coast today was initiated during these earlier highstands, and has been reactivated over the past 6000 years, during the present highstand. Evidence for such prior erosion is found in higher inactive platforms formed during the former Pleistocene highstands (see Figures 1.7 and 7.15) and the wearing down of earlier, higher platforms to a new, lower level. This fact complicates attempts to relate present cliff and platform formation to purely contemporary processes; the rates of erosion indicated above may include some earlier erosion and overestimate the rate of contemporary erosion.

Figure 7.15 *A raised Pleistocene rock platform at Pennington Bay on Kangaroo Island. It is fronted by the modern wave-washed,* active *intertidal rock platform,* both cut into dune calcarenite. Photo: A.D. Short

The secondary role of rocky shores

Rock forms the basement of the entire Australian coast and outcrops, as rocky shore, reefs, islets and islands around approximately 40 per cent of the coast. In addition to providing 12 000 kilometres of rocky shore and environments, the presence of rock has five important secondary effects on the intervening and backing beach systems.

First, the rocky shore exerts considerable control over the length of Australian beaches. By world standards, Australian beaches are short, averaging only 1.4 kilometres in length. This is because most are bordered by some form of rock boundary – a headland, rocks, reefs, or foreland forming in lee of islets and islands. The only three beaches exceeding 100 kilometres are located along the shoreline of active sedimentary basins (see box 'Australia's shortest and longest beaches', page 108). Second, the reduction in wave height that occurs as deepwater waves move into shallower water is attenuated as waves pass over rocky seafloor and reefs. The result is a reduction in wave height at the shoreline, and consequently lower-energy beaches (see Chapter 5).

Third is the impact of wave refraction, caused by waves bending to parallel the seabed contours. As waves approach the coast and encounter rocky seafloor, headlands and islands, they bend, or refract, around these obstacles, particularly those bounded by headlands or rocks. Consequently, Australian beaches tend to be both short and curved, often spiralling towards the lower-energy, more protected end in lee of a headland. Fourth is the impact of the headland, rocks and reefs on circulation in the surf zone. As we discussed in Chapter 5, when any large obstacle is present in the surf zone it will deflect surf-zone currents seaward, forming a topographic-controlled rip (see Figure 5.15). The dominance of short, headland-bound Australian beaches means that when there is surf against the headlands, topographic rips flow out against the boundary, resulting in an average of 4000 topographic rips operating on an average day, particularly around the higher-energy, southern half of the continent.

Finally, during big seas, topographic rips increase in size and may drain an entire embayed beach system, as a megarip. These usually flow out against one or both headlands and well seaward. Such rips occur during high waves on all exposed high-energy, headland-bound beaches, particularly in southern Australia.

In summary, the rocks truncate the shoreline into short beaches, with waves reduced in height by wave attenuation across rocky seabed, resulting in lower-energy beaches. The beaches curve due to wave refraction around reefs, island and headlands. Topographic rips flow out against the rocky boundaries on energetic beaches, and during big seas, headland-controlled megarips drain the entire embayment.

Australia's most spectacular sections of rocky coast

With 12 000 kilometres of rocky shore, it is to be expected that there will be some sections of the Australian coast which – due to a combination of height, rock type, exposure, weathering and accessibility – will attract public attention to the extent they may become major tourist attractions. Some of the more spectacular locations are described below, with a focus on the role natural processes played in their formation. We begin the overview at Cape York, working clockwise around the coast.

Cape York Peninsula

Cape Melville is one of a series of prominent headlands on eastern Cape York Peninsula, located between Cape York and Cooktown. The cape is composed of Permian granite and has been deeply weathered by the tropical climate to expose large granite boulders that form a massive boulder slope, rising to 176 metres. Rainwater falling on the boulders seeps down and emerges at the shore as freshwater springs. Cape Bedford, in contrast, is composed of more resistant, horizontally bedded sandstone, which has remained as a flat-topped, 268-metre-high cape, while the underlying, softer rocks have been eroded to form more gentle slopes (see Figure 7.16).

Lord Howe Island

Lord Howe Island is composed of basaltic volcanic rock, but with parts having a veneer of dune calcarenite. These two lithologies serve to demonstrate the important role that rock resistance plays in determining coastal morphology. The spectacular, near-vertical cliffs along the southern shores of Lord Howe Island are formed from the horizontally bedded Mount Lidgbird basalt, and contrast with the less resistant aeolianites, which have a much more subdued topography.

Figure 7.16 *Flat-topped Cape Bedford, at Cape York Peninsula in Queensland, is formed of horizontally bedded sandstone and rises to 268 metres.* Photo: A.D. Short

The island shores are also exposed to different wave energies – from those exposed to the full force of the Tasman Sea to the western shore that is sheltered behind a fringing reef. The reef attenuates wave energy substantially and there is a marked contrast between the exposed, south-facing, plunging basalt cliffs of Mount Gower and Mount Lidgbird, compared to gentle hill slopes formed on similar basaltic rocks, where they back the sheltered lagoon (see Figure 8.7a). Likewise, calcarenite cliffs on exposed, east-facing calcarenite shores erode relatively rapidly and develop sub-horizontal, mid-tide rock platforms, which can be several hundred metres wide. Their low elevation is a result of their porous nature and low level of saturation. But in similar locations in the more resistant basalt, platforms are absent or narrower and at a higher level, reflecting a slower rate of cliff retreat and a higher level of saturation. The most impressive cliffs on Lord Howe Island, and the highest cliffs in Australia, occur around the southern peaks, where cliffs are essentially vertical for 500 to 800 metres and then plunge into water depths of between 5 and 20 metres without a clear change in their profile at sea level.

Sydney Coast

The Sydney coast is renowned for its beaches, yet 75 per cent of the coast is rocky. Three factors determine the form of the coast. First is the massive, horizontally bedded sandstone and thinner shale beds, which have two major joint lines orientated roughly north–south and east–west. The joint lines determine both the direction of river flow and orientation of the cliff lines. Both the bedding and joints provide lines of weakness and in part determine the shape and size of the sandstone boulders that occasionally fall from the cliffs, and those quarried by wave attack. Second is the general height of the cliff line, from 20 metres to as high as 100 metres at North Head, causing them to stand proud of the shoreline and to form prominent headlands bordering the beaches.

Third, the two rock types, sandstone and shale, produce contrasting landforms. The resistant, massive sandstone forms vertical cliffs, which dominate much of the coast and frame the entrance to Sydney Harbour at North Head and south of South Head (see Figure 7.4), while the weaker shales result in a sloping surface, such as that at Long Reef Point. At the base of the cliffs are rock platforms formed in both the sandstone, as relatively narrow high platforms, and in the softer shale where they form wider, lower platforms; both support rich intertidal and subtidal ecosystems.

Great Ocean Road

The 260-kilometre-long Great Ocean Road, located between Anglesea and Warrnambool in Victoria, is truly one of the great coastal drives in the world. The main reason is that it traverses a predominantly steep rocky coast that provides spectacular views, and culminates in the rapidly eroding limestone and marls of the Port Campbell National Park, which is best exemplified by the Twelve Apostles. The coast consists of two parts, the eastern Anglesea to Cape Otway section, where the road hugs the coast and which is composed of Jurassic-age sedimentary rocks, including horizontally bedded sandstone, siltstones and conglomerates, and the western Cape Otway to Warrnambool section, which is predominantly Tertiary limestone and marls.

On the eastern Anglesea coast, the sedimentary rocks are eroded from prominent rock platforms backed by steep cliffs, and cut by occasionally narrow, V-shaped valleys. The road winds around the cliffs and into and out of the valleys. The softer limestone of the western Port Campbell coast, which bears the full force of the Southern Ocean swell, is eroding far more rapidly (1–20 millimetres per year on average) forming steep, at times overhanging, cliffs up to 100 metres high. The rapid retreat also leaves sea stacks (see Figure 7.17) such as the Twelve Apostles and arches such as the former London Bridge. Both of these iconic landforms have undergone major collapses in the past 20 years. London Bridge comprised a double arch, but the inner arch collapsed on 15 January 1990. Similarly, the collapse of one of the sea stacks within the Twelve Apostles occurred on 3 July 2005, resulting in fewer 'apostles' left standing.

Tasmania

Tasmania has a number of notable rocky coastal features. At the Tessellated Pavement, located on the Tasman Peninsula, near-horizontal siltstone is dissected by a series of joints. These joints are aligned so that the surface has been sculpted into a series of 'tiles', through salt crystallisation and removal of weathered material by wave action. Tasmania's most photogenic, highest, but also most inaccessible, cliffs are the 300-metre-high columns at Cape Pillar (see Figure 7.18), Raoul Head (230 metres) and the 272-metre-high Fluted Cape on Bruny Island. All are composed of the Jurassic dolerite that dominates the entire east and south coast between Swansea and South Cape. At Cape Pillar, the rocks are both high and eroded to expose the tall columns of dolerite, stacked like organ pipes.

The Nut, on the north coast at Stanley, is a 143-metre-high, flat-topped volcanic plug, formed of Tertiary basalt. The surrounding volcano has been eroded to leave the more resilient central plug of the volcano. Today, it stands proud of the shore, linked to the mainland by long curving beaches to either side, with the town of Stanley nestled on its southern base. Table Cape, near Wynyard, and The Bluff at Devonport are of similar origin.

South Australia

The most common rock type found across the southern coast is calcarenite, lithified Pleistocene sand dunes that blanket older rocks and deposits, and now form reefs, islets, islands, headlands and cliffs along hundreds of kilometres of the coast (see Chapter 6 and Figure 7.19). This rock type reaches

Figure 7.17 Sea stacks at Dog Trap Bay, Victoria. Photo: A.D. Short

Figure 7.18 Cape Pillar is named after its prominent dolerite columns. Photo: A.D. Short

Figure 7.19 Calcarenite along the southern Australia coast: **a)** Young incipient calcrete is scaped foredune showing friable root casts (Thomas River, Western Australia); **b)** A calcarenite cliff showing old-soil horizons (horizontal structures) and lithified dune bedding (sloping structures), Robe, South Australia. Photos: A.D. Short

its greatest extent and influence in South Australia, particularly along the 45-kilometre-long Canunda National Park and the 45-kilometre-long Robe Range in the southeast; the 235-kilometre-long south and west coast of Kangaroo Island; and along the 1130-kilometre-long western Eyre Peninsula coast.

Western Australia

In Western Australia, calcarenite occurs along much of the coast between Cape Leeuwin and Shark Bay, with the best example near Perth being Rottnest Island, entirely formed of calcarenite. The greatest extent is on the central Western Australia coast, along the 180-kilometre-long, 100–200-metre-high Zuytdorp Cliffs (see box 'Zuytdorp Cliffs', page 184) and the adjoining Dirk Hartog and Bernier islands. In all areas, the calcarenite forms cliffs, with each cliff consisting of multiple layers of buried former soils (see Figure 6.23) up to 200 metres high. The weakly consolidated calcarenite erodes relatively rapidly, leaving vertical cliffs that are usually fronted by mid-tide horizontal rock platforms, along with rocks, reefs, islets and islands. The best locations to observe the calcarenite formations are at Robe, Whalers Way at the southern tip of the Eyre Peninsula, and at Elliston, all in South Australia.

The Bunda cliffs, along the southern edge of the Nullarbor Plain, is the longest section of continuous coastal cliffs in Australia. They extend for 210 kilometres between Head of Bight and Eucla, with an average height of 60–90 metres (see Figure 7.20). They then continue up to 40 kilometres inland as the Hampton Bluffs, for 300 kilometres to Twilight Cove, then are again exposed along the coast as the 100–120-metre-high Baxter Cliffs, which extend for another 160 kilometres to Point Culver, before again tending inland as the Wylie escarpment for 120-kilometres. In total, the continuous cliffs and bluffs that form the southern boundary of the Nullarbor Plain occupy 790 kilometres of the Great Australian Bight, the longest cliff line in the world. Baxter and Wylie were John Eyre's companions on his historic walk across the Bight in 1841, and Baxter monument marks the remote location where Baxter was killed. It is one of the few human landmarks on the entire Bight coast, a long, lonely section of dramatic coast that has changed little since Eyre's trek.

The south coast of Western Australia, between the low Point Lorenzen at the western end of the Great Australian Bight and the conical Wylie Head just east of Esperance, marks the boundary of a 300-kilometre-long section of coast dominated by massive Proterozoic granite. So spectacular is the coast that 73 per cent of it is reserved, from the east, in the Nuytsland Nature Reserve and the Cape Arid and Cape Le Grande national parks. The granite is noteworthy for several reasons. First, it is a resistant, impermeable rock that usually slopes steeply into the sea, without any obvious marine erosion other than exposure of the bare rock. Second, the same rock outcrops along the coastal hinterland as high granite domes, called 'tors', and offshore as the 105 rounded islands and numerous reefs of the Recherche Archipelago.

Figure 7.20 *The 90-metre-high Nullarbor cliffs in South Australia, exposing the upper, reddish Nullarbor Limestone and lower, whitish Wilson Bluff Limestone. The cliffs extend continuously for 210 kilometres between Head of Bight and Wilsons Bluff, on the South Australia–Western Australian border.* Photo: A.D. Short

Figure 7.21 *Cape Leveque is composed of horizontally bedded white sandstone, which has been deeply weathered to form the bright red pindan soils.* Photo: A.D. Short

Third, as the regional granite has eroded over time it has delivered quartz sand to the coast, which has formed most of the beaches, the majority composed of pure white quartz. The persistent Southern Ocean swell breaking in crystal-clear water over the white sand, bounded by the sloping granite and granite boulders provides some of the most scenic and pristine coastal sights in Australia (see Figure 5.34). Similar granite also outcrops along 280 kilometres of the South Coast, between Lookout Point and Cliffy Head, and on the 115-kilometre-long Leeuwin Coast, between Cape Leeuwin and Cape Naturaliste, where much of it is blanketed by dune calcarenite.

Cape Leveque, north of Broome, has become one of the more photographed locations on the Australian coast, owing to its deeply weathered, brilliant-red, horizontally bedded Cretaceous sandstones. If you follow the 300 kilometres of coast between Broome and the cape, the most distinctive feature is the bright red of the backing low-pindan plain, which is usually eroded and forms cliffs a few metres high along the coast (see Figure 7.21). The pindan is a deep red soil. At the cape, the plain rises to a height of 30 metres and erosion around the base of the cape has exposed both the pindan and underlying, horizontally bedded red sandstone, which extends offshore as intertidal rocks and subtidal reefs. The red rocks are highlighted by

Figure 7.22 a) The flat-topped, 50-metre-high Cape Bougainville shows part of the original plateau surface and is also formed from Proterozoic basalt. b) Two joint-controlled valleys and their beaches, near Buckle Head in the east Kimberley. Photos: A.D. Short

white beach sand, deposited around the base of the rocks, while immediately to the east of the cape the white sand dunes intermingle with the red pindan to produce a range of colours, from white to pink to red.

The Kimberley coast, between Point Usborne at the eastern entrance to King Sound and Cape Dussejour at the western entrance to Cambridge Gulf, is the longest section of rocky coast in Australia, with 2565 kilometres of open coast, 80 per cent of which is rock. In addition, many of the 1181 beaches along this section, which average only 300 metres in length, are bordered and backed by rocky points, bluffs and cliffs, ranging between 20-100 metres high at the coast. The Kimberley coast has just two types of rocks, the predominantly horizontally bedded, red sedimentary sandstone, siltstone and conglomerates deposited in a basin environment up to 2000 million years ago, and the basalt that flowed onto the then-seafloor 1800 million years ago (see Figure 7.22a). The ancient sediments and basalts were uplifted to form a flat central plateau, which was deeply jointed along roughly north–south and east–west orientations. The joints predetermine the direction of river and creek flow, which over many hundreds of millions of years had denuded the plateau to erode deep straight valleys (see Figure 7.22b) in the plateau, many interrupted by waterfalls as well as numerous cliffs.

Figure 7.23 a) Eroding laterite bluffs at Cape Charles, fronted by intertidal laterite reefs. *b)* The highest rocky coast in the Northern Territory is the dipping, 40 metre-high sandstone cliffs at Cape Wilberforce. Photos: A.D. Short

Added to this is the deep tropical weathering associated with Tertiary laterite soil formation, which in many areas results in a red surface layer over a thick white 'pallid' soil horizon. This is underlain by a zone of red nodule accumulation. Finally, coral reefs fringe both the rocky shore and 163 of the beaches. When all of these are combined in a tropical climate that permits prolific mangrove growth in sheltered areas, in an almost uninhabited terrain and along with the occasional boab tree and the ever-present crocodiles, it makes for one of the most beautiful and interesting stretches of the world's coastline.

Northern Territory

For the most part, the Northern Territory coast is dominated by sedimentary basins, resulting in low sections of rocky shore, rarely rising above a few metres. Like the Kimberley coast, the bedrock has been deeply weathered to form laterite which, when eroded, is exposed as bright red and white bluffs. As it erodes it forms intertidal and subtidal laterite reefs (see Figure 7.23a). The highest sections of rocky coast on the mainland occur at Cape Wilberforce, where Proterozoic sandstones rise to 40 metres. On the eastern side of the cape, the cliffs extend for 20 kilometres to the southwest, the longest rocky section of the Territory coast (see Figure 7.23b). On either side of the Gulf of Carpentaria, centred on Cape Arnhem in the west and Weipa in the east, the horizontally bedded sedimentary rocks have also been deeply weathered to form laterite. Laterite is a humid tropical soil whose formation involves the precipitation of nodules of magnesium and aluminium, which can result in deposits called bauxite. The bauxite – which is the raw material for aluminium – is mined at Gove in Arnhem Land, on Groote Eylandt and at Weipa. Northern Australia has 40 per cent of the world's bauxite reserves.

Conclusion

Australia has an extensive and varied rocky coast. Rock types include the igneous granites and basalts, the full range of sedimentary rocks including extensive areas of calcarenite in the south and west, and large areas of the resistant metasedimentary rocks. Present and past sea levels have attacked these rocky shores to form a range of features, from the intact, resistant, usually sloping granites, to the cliffed basalts and sandstones, usually fronted by rock platforms, and the etched and often jagged metasedimentary rocks. In places, the rocks form spectacular landscapes, some of which have become major tourist attractions. While marine attack has eroded and etched the rocky coasts, they represent by far the most resistant and usually the higher sections of the Australian coastline, and in a time of accelerating sea-level rise will provide a higher and more stable base to much of the Australian coast.

REEF COASTS

Introduction

Coral reefs are remarkable ecosystems; they inspire a sense of wonder, whether you are flying over them or diving amongst them. As we shall see, they impart a distinctive character to the shoreline. Corals form the framework for a limestone structure that modifies the wave energy reaching the coast and provides carbonate sediments from which beaches and islands are built.

Australia has the most extensive and pristine reef systems in the world. The northeast coast is dominated by the Great Barrier Reef, which extends for 2300 kilometres. Australian waters also include other extensive but lesser-known reefs. Isolated reefs rise from the continental shelves beneath the Timor and Arafura seas. Reefs fringe many of the headlands along the coast of northern Australia, and Ningaloo Reef, along the northwest coast of Western Australia, is the largest fringing reef in the world. The southernmost reef in the Pacific occurs at Lord Howe Island (31 degrees 30 minutes South), whereas the southernmost reefs in the Indian Ocean are at the Abrolhos Islands (29 degrees South). There is only one true atoll – the Cocos (Keeling) Islands, an Australian territory in the Indian Ocean – but it has played a central role in our understanding of reef development. There are isolated reefs in the Coral and northern Tasman seas, and non-reef-forming coral communities close to the mainland coast at the Solitary Islands, near Coffs Harbour (30 degrees South) and on Rottnest Island (32 degrees South).

In Chapter 7 we described the rocky coasts and we saw how the persistent forces of the sea gradually erode and shape the rocky shoreline, whether it is composed of highly resistant, crystalline

rocks or poorly lithified, fossil dune calcarenite. In this chapter, we consider the remarkable role that coral reefs play in reducing wave energy and protecting the shore from marine attack.

We begin with a description of tropical and subtropical reefs, exploring how the living coral covers a limestone structure. Corals form the framework that gradually, almost imperceptibly, builds a reef – a recent veneer over older reef limestone – punctuated by a series of geological events, such as sea-level fluctuations.

How coral reefs develop

Coral reefs flourish where wave energy is high, and are better developed on exposed, windward shores than sheltered, leeward shores. Reefs prefer clear waters, with corals extending beyond depths of 30 metres, where light can penetrate that deep. They also require a suitable foundation on which to establish, and often reoccupy former reef positions. The best-developed reefs are therefore found growing on older reef structures, in clear water that is exposed to ocean swell.

The broad geographical distribution of reefs is also influenced by factors such as water temperature, salinity and substrate availability and stability. In addition, regional patterns of currents and waves influence larval dispersal and recruitment. The distribution of species and particular growth morphologies across an individual reef is dependent upon more subtle gradients in environmental factors, including water depth, which influences the amount of light that is available for corals to survive, and sediment transport, which is also a function of wave energy.

Corals are constrained by salinity. They do not establish where there is a large freshwater input, such as at the mouths of large rivers that also carry increased sediment or nutrient loads. Reefs appear to flourish where water temperatures are favourable, but in fact corals are limited by both lower and upper temperature thresholds. Corals are rare in waters where average monthly sea surface temperatures fall below 18 degrees Celsius. Corals are also sensitive to excessively high temperatures; they can tolerate temperatures of up to around 30 degrees Celsius, but if the seasonal maximum to which they are accustomed is exceeded by 1 to 2 degrees, they are subject to bleaching. Bleaching occurs when corals expel their symbiotic algae, called zooxanthellae, thus losing their colour and appearing white. Although reefs appear to be robust, forming a rigid limestone structure that persists in the geological record, there is much evidence that the corals themselves are fragile and sensitive to a series of stresses. Recent episodes of mass bleaching on reefs around Australia appear to be a result of global warming, and are a major cause for alarm in terms of the impacts of future climate change, described in more detail in Chapter 9.

Reef structure

When swimming over a reef it is clear that corals cover much of the surface and provide habitat for a great diversity of other marine organisms, from cryptic invertebrates to vast schools of colourful fish. It may be less clear beneath the waters, but is apparent from the outcrops of fossilised reef on land, that the skeletal remains of the corals themselves

Figure 8.1 *A view of the island of Pulu Pandan, on the eastern rim of the Cocos (Keeling) Islands. The reef crest is clearly visible as a 'hardline' along which waves are breaking. The forereef is characterised by a distinct spur and groove at right angles to the reef front. The backreef is a shallow lagoon, in some cases terminating in the island but elsewhere grading into the inter-island passages.* Photo: C.D. Woodroffe

form the primary framework of the reef limestone. The framework is partially lithified (turned into rock) by secondary builders, particularly coralline algae that encrust the surface and give it further rigidity. However, reefs are also subject to physical, biological and chemical processes of destruction. For example, stormwaves break fragile corals, and their surface is continually grazed and bored by organisms such as parrot fish, which rasp the coral, or molluscs and sponges that bore into the limestone. Sediment from this breakdown and bio-erosion of reef limestone, as well as from the skeletal remains of other smaller organisms, such as foraminifera, provides a rain of fine sand that accumulates in voids and gradually becomes cemented through chemical processes. Reefs are built only where the net accumulation of carbonate exceeds the processes of destruction.

Reef morphology

The massive limestone structures that form coral reefs can be divided into three principal environments: forereef, reef crest and backreef (see Figure 8.1).

Figure 8.2 *A view of the algal rim on the reef crest at Home Island, Cocos (Keeling) Islands, showing the pink coralline algae that cover the reef crest and the reef flat that largely dries at low tide and fronts the island.* Photo: C.D. Woodroffe

Reef crest

The reef crest is especially prominent, and is marked by a 'hardline' at which waves break. Corals may grow on the reef crest in low-energy settings, but encrusting coralline algae are better able to tolerate the force of the waves breaking and exposure during low tide, and more often the reef crest is dominated by an algal rim (see Figure 8.2). Waves break on the reef crest, losing 70–90 per cent of their energy, and the backreef is considerably more sheltered, although some waves may reform to cross the reef flat at high tide and enter the backreef area.

Forereef

To seaward of the reef crest, the forereef generally rises steeply from deep water. When exposed to powerful swell, the reef front often contains a prominent spur and groove topography that is at right angles to the reef crest. Waves approaching the reef undergo little attenuation, or refraction, prior to reaching reefs, until they encounter the coral-covered spurs. They then break heavily on the reef crest as plunging waves, such as can be seen at the famous Teahupoo surf break in Tahiti.

Backreef

The backreef environment, behind the reef crest, can form either a shallow lagoon or a reef flat. A reef flat is a near-horizontal platform, often several hundred metres wide, that is flooded at high tide but retains little water over it at low tide, in some cases draining to leave an emergent surface. This is particularly the case on atolls, such as the Cocos (Keeling) Islands (see Figure 8.2), and on platform reefs such as those in Torres Strait, where the reef flat was formed when sea level was slightly higher. Corals cannot tolerate more than the very briefest exposure above the water surface at lowest tides, and so do not occur on emergent reef-flat surfaces. Emergent reef flats are usually relatively bare of sediment, whereas shallow lagoons often contain isolated patch reefs around which sediment accumulates. In the case of the Great Barrier Reef, there is a relatively deep lagoon behind the outer reef, described later in this chapter, within which smaller platform reefs themselves contain lagoons.

Whereas the reef structure is built through the binding of corals and coralline algae, carbonate sediments are produced by bio-erosion and from the skeletons of the smaller calcareous organisms, particularly foraminifera, that live within the reef. Massive corals can be detached by tropical cyclones, and then form large reef blocks or boulders. Branching corals fracture into finger-sized sediments, called shingle, that can be swept into ridges by storms, and the efficacy of bio-erosion and rapid growth of the small calcareous organisms mean that there is abundant sand-sized sediment on a reef. The rate of sediment production on reefs is so rapid (see box 'Coral growth and reef formation', right) that carbonate sediment dominates the shores

CORAL GROWTH AND REEF FORMATION

Individual coral colonies can adopt a variety of growth forms, depending on the genus of coral and the environment in which it grows. Individual coral colonies grow by calcification – the coral polyps secrete carbonate that forms the underlying skeleton (exoskeleton) on which the polyps grow. Growth forms vary from massive spherical colonies of the genus *Porites* with skeletal extension rates of 10–20 millimetres per year, to delicately branching forms of the genus *Acropora* that can extend at 100 millimetres per year or more. The rate of reef formation, however, is only a few millimetres a year, much slower than the rate of growth of the individual coral colonies, because the reef is the net aggregate of vertical frame accumulation, encrustation and sediment infill. Reefs are also being eroded and tend to be extremely porous, with many unfilled cavities. The rate at which carbonate accumulation occurs on a reef can be assessed by at least three independent methods: coring and radiometrically dating a reef to determine its growth over past centuries, estimating the abundance of calcifying organisms and multiplying by their inferred rate of growth, or measuring the chemical balance of waters moving across the reef to infer carbonate production as alkalinity changes. These different methods have produced broadly similar rates of reef growth or sediment accumulation, implying that the active reef accumulates carbonate at around 10 kilograms per square metre per year (a vertical growth potential of around 7 millimetres per year). Coral patches accumulate carbonate at around 4 kilograms per square metre per year (a vertical growth potential of around 3 millimetres per year). Sand-covered reef flat and lagoon floor accumulates carbonate at <0.5 kilogram per square metre per year (a vertical growth potential of around <0.2 millimetres per year).

of high volcanic islands as well as the low cays away from the mainland (see Figure 8.3). However, along the mainland coast, calcareous sediments are diluted by terrestrial sediments that are delivered by rivers and from weathering of adjacent rocky shores. There is generally a gradation in the carbonate proportion across the reef, from areas that are entirely carbonate to those with relatively little local carbonate production to dilute the quartz-rich sands derived from coastal rivers and creeks (see Figure 8.3).

Figure 8.3 *The proportion of carbonate sands that accumulate depends on the ratio between local carbonate production and supply of terrestrial sediment.* ***a)*** *Beaches backing a fringing reef on the Kimberley coast are more than 50 per cent carbonate, despite the cliffs behind them.* Photo: A.D. Short ***b)*** *The significance of the fringing reef can be seen on Dawar, one of the Murray Islands in Torres Strait. Despite the volcanic nature of the island, the spit and beach that have accumulated are composed principally of carbonate sand.* Photo: J. Leon

Oceanographic processes and their effect on reefs

Reefs flourish where wave energy is high, but the structure of the reefs also influences the distribution of wave and tidal energy. River processes, together with sediment and nutrient loads, exert further constraints on reef development along continental margins. The surface morphology of a reef reflects its adjustment to the contemporary processes that shape it, as well as the stage of evolution that an individual reef has reached. Reefs also adjust to changes, particularly of sea level, over decades and centuries.

Wave processes

Wave energy exerts considerable control on both reef morphology and ecology. Exposed windward reef crests on mid-ocean reefs, like the Cocos (Keeling) Islands, have well-developed reef crests on the southeastern margin, where they face the dominant southeast wind and swell. Similarly, the outer reefs of the Great Barrier Reef are swell-dominated, with steep, plunging breakers averaging 1 to 1.5 metres high and with periods of 8 to 10 seconds, while the sheltered Great Barrier Reef lagoon is dominated by smaller, locally generated and fetch-limited wind waves whose height is <0.5 metre and period is generally 3 to 4 seconds. As considerable wave energy is dissipated on the reef crest, the backreef environments consequently receive substantially less wave energy. The leeward reefs tend to be less abrupt in slope and usually contain a greater diversity of coral. On individual reef platforms, waves wrap around and break on all sides of the reef, including its leeward margin.

Wave activity is also tide-dependent, particularly where the reef flat dries at low tide. During the rising tide, these emergent reef flats are progressively inundated and waves increase in size as water depth increases. Wave height, during high tide, depends on the depth of water, but decreases with distance across the reef flats as a result of bottom friction. However, persistent winds can cause locally generated, short-period waves that are refracted to converge towards the lee of the platform. These waves can transport sediment towards the lee of the platform, and reef islands are typically found on the leeward of such platforms where waves converge as a result of refraction.

Reefs exposed to high wave-energy, such as most of the outer reefs of the Great Barrier Reef, are swept clean of sediment and rarely develop reef islands. Considerable sediment movement can occur during tropical cyclones (see box 'Tropical cyclones on the Great Barrier Reef', right) when these high-energy storm events can pile up coral debris into ridges along the windward margin.

Tidal and river processes

Tides exert several controls on reefs. The efficiency of wave processes depends on the stage of the tide, as well as reef elevation and the width of the backreef environment. The tidal range on the Great Barrier Reef is large compared with mid-ocean reefs where it is usually less than 1 metre. The tidal range exceeds 6 metres in the vicinity of Broad Sound (22 degrees South), where reefs are absent because tidal currents entrain mud, making it too turbid for coral larvae to establish. Tidal currents in excess of 2 metres per second are common in parts of Torres Strait because of its position between the tidal systems of the Coral Sea and those of the Gulf of Carpentaria.

Strong tidal currents can influence reef shape, and many of the reef platforms in Torres Strait are linear with a west–east orientation, parallel to these strong tidal flows. Tidal currents are also accentuated where reef density amplifies the flows; for example, narrow passages through the outer reef crest experience tidal flows of up to 4 metres per second, giving rise to tidally formed 'deltaic' reefs, where the outer reef is discontinuous in the northern Great Barrier Reef.

TROPICAL CYCLONES ON THE GREAT BARRIER REEF

The Great Barrier Reef is subject to occasional, tropical cyclone-generated waves. More than 80 cyclones have been recorded south of 10 degrees South in the past 40 years (see Figure 2.3) of which about 10 per cent were category 3 or above, and some, such as Cyclone Ingrid in 2005, were in the highest category, 5. Tropical cyclones cause a range of damage to coral reefs. The most common damage includes broken and dislodged corals, which are moved and deposited as coral rubble. Massive coral heads, called 'bommies', can be toppled by the largest waves, and sheltered backreef environments may also sustain damage to the more delicate branching corals that then contribute to the formation of shingle ridges, which are prominent on the margins of some reef platforms.

Strong tidal currents also characterise the passages of the Pompey complex of reefs (20–21 degrees South).

River influence exerts a negative control on reef establishment through freshwater and sediment load, as corals cannot tolerate freshwater, or blanketing by sediment. Nearshore reefs tend to occur in the upper 5 metres, and corals may be smothered if sediment accumulation there is too rapid. For this reason, reefs are absent along much of the Queensland mainland, particularly where major rivers reach the coast.

The impact of sediment carried by the rivers may, however, be overstated, as some reefs can survive in quite turbid conditions. There are more than 100 nearshore reefs, such as those at Paluma Shoals, and they are quite diverse. Large floods, associated with tropical cyclones, result in substantial sediment plumes, in some cases reaching mid-shelf reefs, and in exceptional circumstances even to the outer reefs, which can adversely impact the Great Barrier Reef well offshore. Along the northeastern Queensland coast, mud is often concentrated in sheltered, northward-facing bays and gulfs that are protected from the prevailing southeasterly winds. These are also the favoured sites for mangrove colonisation and are usually bare of corals.

Reef types and geological evolution

Charles Darwin proposed a theory of reef development based on the interrelationship between fringing reefs, barrier reefs and atolls. Darwin envisaged that each might represent a different stage in an evolutionary sequence. Having witnessed uplift following an earthquake on the coast of South America, he postulated that a volcano in mid-ocean might subside, but that a surrounding reef would be able to remain at sea level as a result of vertical reef growth. An initial fringing reef would develop into a barrier reef, with a lagoon forming behind the reef crest. If subsidence continued still further, this would lead to the total submergence of the volcano, and an atoll would result, with reef growth maintaining the rim at sea level (see Figure 8.4).

The only atolls that Darwin visited were the Cocos (Keeling) Islands, and these became central to his argument. Subsequent drilling on several Pacific atolls has shown that volcanic rocks do underlie atolls, often at depths of a kilometre or more, which provides support for Darwin's theory. The distribution of oceanic reefs is now explained in terms of plate tectonics, and several stages of the Darwinian model occur along linear island chains, such as the Hawaiian Islands, where older islands are at progressively more advanced stages of development. Whereas this explains the gross structure, the surface morphology of reefs – and hence the habitats that now characterise those reefs – is better explained in terms of the reef's response to past changes of sea level. Cores through reef limestone indicate that reefs were exposed during lowstands of sea level, with reef growth resuming when sea level again rose during interglacials, as we discuss below.

Open-ocean reefs

Cocos (Keeling) Islands

The Cocos (Keeling) Islands, located in the eastern Indian Ocean (12 degrees South, 97 degrees East) are the only Australian example of a true atoll. In contrast to archipelagos of atolls elsewhere in the Indian and Pacific oceans, this group of islands is isolated and is not part of a chain of volcanoes. The group comprises the South Keeling Islands, which form a horseshoe-shaped atoll (see Figure 8.5), and a single island, North Keeling, 27 kilometres to the north. There has been only shallow drilling on the Cocos (Keeling) Islands since Darwin and several subsequent naturalists visited the islands. Undoubtedly, volcanic basement underlies this atoll, but its depth remains unknown. The shallow drilling has encountered an older Pleistocene reef limestone, over which the modern reef rim has formed.

Figure 8.6 shows a typical cross-section across the reef rim on the Cocos (Keeling) Islands. The reef front is sculpted into a sequence of grooves and spurs (see Figure 8.1), with a prominent algal rim at the reef crest (see Figure 8.2), and wave energy rapidly decreases across the near-horizontal reef flat. The morphology of islands is also shown in Figure 8.6. Prominent on the windward shore, around the main rim, is a coral conglomerate, forming a near-horizontal platform composed of cemented boulders of massive coral and disoriented branches of smaller corals eroded from the reef front. Radiocarbon dating indicates that the platform formed as a reef flat, slightly above the level of the modern reef flat, around 3000 to 4000 years ago, indicating a higher sea level at that time. This provides a foundation on which the sandy reef islands have accumulated.

Islands typically consist of a steep windward shore, rising to an elevation of around 3 metres above mean sea level, behind which is a lower-lying central

Fringing reef

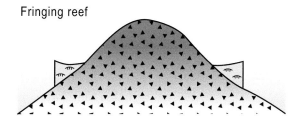

Barrier reef

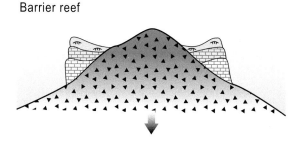

Atoll reef

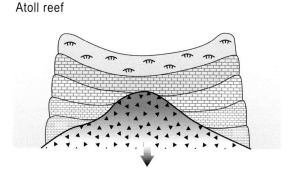

Figure 8.4 A schematic illustration of fringing reef, barrier reef and atoll reef, the three stages envisaged by Darwin in an evolutionary sequence of open-ocean reef types.

217

Figure 8.5 *An open-ocean reef, the southern margin of the Cocos (Keeling) Islands, an Australian atoll in the Indian Ocean.* Photo: C.D. Woodroffe

area, in some cases with a less-distinct ridge on the lagoon shore. Islands are predominantly sandy, with shingle in some places and coral boulders along the windward shore of islands on the eastern rim, where the reef flat is narrowest. The schematic diagram shows that the degree of cementation decreases with depth and with distance across the reef island, and that a freshwater lens occurs beneath the island. It also indicates that, at a depth of 10 to 15 metres below sea level, there is an older Pleistocene reef limestone, which has been shown by radiometric dating to be of Last Interglacial age, on which the modern reef rim has formed.

Lord Howe Island

Lord Howe Island is a volcanic island with a partial fringing reef. It contrasts with two atoll-like reefs further north in a foreshortened chain of islands in the Tasman Sea, migrating gradually northwards on the Indo-Australian plate and which provide further insights into stages of reef development in mid-ocean. At the southern end of the chain lies Balls Pyramid (31 degrees 46 minutes South), an impressive, reef-less volcanic pinnacle that rises to 551 metres. It sits in the middle of a shelf that is at least 15 kilometres wide, represents the penultimate stage in the truncation of a volcano and indicates the effectiveness of marine erosion. Lord Howe Island (31 degrees 30 minutes South), a crescent-shaped, Late Miocene volcanic island, demonstrates the next stage (see Figure 8.7a). It occurs in the middle of a shelf of around 50 metres water depth that is 24 kilometres wide and 36 kilometres from north to south. The island is dominated by the peaks of Mount Gower (875 metres) and Mount Lidgbird (777 metres), each of which is flanked by steep, plunging cliffs. However, along the western shore is a fringing reef, the southernmost coral reef in the Pacific Ocean. In contrast to the precipitous cliffs on the exposed shores, the reef reduces wave energy and encloses a shallow lagoon whose inner shore is composed of rounded hill slopes and a long sandy beach.

There are two atoll-like reefs more than 100 kilometres to the north of Lord Howe Island. Elizabeth Reef (29 degrees 56 minutes South) is oval-shaped, 10.7 kilometres long and 6.2 kilometres wide. Middleton Reef (29 degrees 27 minutes South) is kidney-shaped, 9.3 kilometres long and 5.7 kilometres wide (see Figure 8.7b). It is unclear whether Elizabeth and Middleton reefs are true coral atolls that have formed as a result of subsidence of the volcanic basement on which they are founded (as suggested in the Darwinian model in Figure 8.4), or whether they represent reefs that have formed on a horizontal truncated shelf, like those around Lord Howe Island (see Figure 8.8). These reefs have moved gradually north as the plate has migrated. To their north, there are submerged banks and guyots, which are underwater seamounts with flat tops that generally are truncated by marine erosion. The guyots represent still-older volcanic

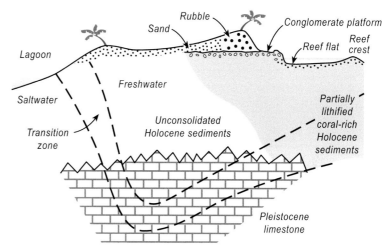

Figure 8.6 *A schematic cross-section of the rim of an atoll, based on studies of the Cocos (Keeling) Islands*

edifices in a seamount chain that is being carried northward as the Australian plate moves (see Chapter 1).

Continental-shelf reef systems

The Great Barrier Reef is a continental-shelf reef system. The Great Barrier Reef can be considered 'great' for several reasons: it is the most extensive reef complex in the world, it can be seen from space and is considered the largest biologically built structure in the world, and it is managed sustainably as part of the world's largest marine park. There are more than 3600 individual reefs that make up the Great Barrier Reef, including more than 750 fringing reefs, which occur along the mainland coast and on 'high' islands where bedrock outcrops.

In the description that follows, the broad structure of the shelf on which the reef has formed and the latitudinal variation in the morphology of

Figure 8.7 a) The fringing reef on the western margin of Lord Howe Island is the southernmost in the Pacific Ocean. Around much of the margin of Lord Howe there are precipitous cliffs such as those flanking Mount Gower (right) and Mount Lidgbird in the background, but where the reef provides protection a sheltered lagoon exists and the hills are gently sloping, which is typical of those on which processes of hill-slope erosion dominate. Photo: C.D. Woodroffe *b) Middleton Reef, a kidney-shaped reef 150 kilometres north of Lord Howe Island, resembles an atoll in form, but may have formed over a truncated platform like that around Lord Howe Island.* Photo: M. Hallam

the outer reefs are described first. Fringing reefs are an important reef type, both on the barrier reef and elsewhere around the coast of Australia; these include incipient fringing reefs and nearshore reefs, and are examined next. Finally, there are also latitudinal differences in the morphology of the reefs that occur in mid-shelf, called 'platform reefs', and these are described in relation to their stage of evolution and infill.

We saw in Chapter 3 that reef ecosystems reach their greatest diversity in northern Australia, with several hundred species of corals. The gradual northwards movement of the Australian plate has carried the continent relentlessly towards this Indo-Pacific centre of diversity. Although there are changes in the nature of the reef from north to south along the length of the Great Barrier Reef, there is a much more significant west–east environmental gradient between the shoreline and outer reef, reflecting the varying influence from terrestrial sources at the coast to the open ocean (see Figure 8.9). On its seaward side, the Great Barrier Reef is influenced by the East Australian current, but circulation across the reef is driven primarily by the southeast trade winds, and complex tidal flows and eddy patterns occur in response to the dense reef networks. Terrestrial sediment, derived from the land, dominates the seafloor close to the mainland, while the mid-shelf is composed of both terrestrial and carbonate sediment, and the outer shelf is more than 80 per cent carbonate.

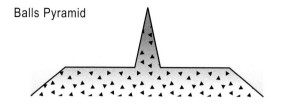

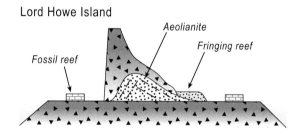

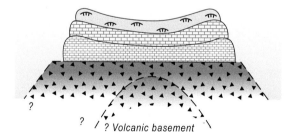

Figure 8.8 *A schematic representation of the Lord Howe seamount chain. Balls Pyramid is a single pinnacle on a largely truncated platform. Lord Howe Island sits on a similar platform, but with fossil dunes and reef (modern fringing reef and fossil reefs on the platform). Middleton and Elizabeth Reefs are atoll-like reefs.*

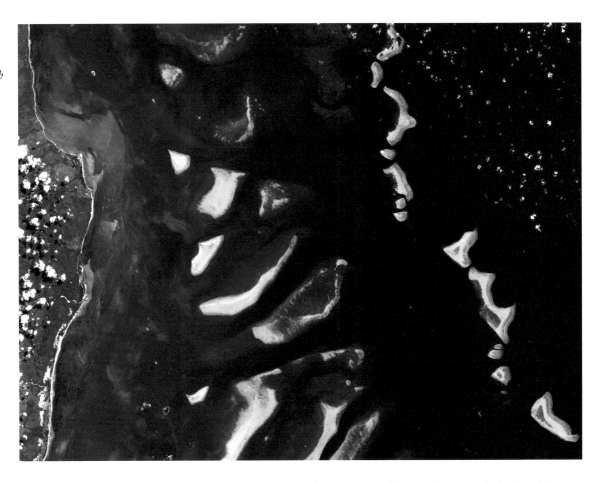

Figure 8.9 *A Landsat satellite image showing part of the northern Great Barrier Reef, north of Princess Charlotte Bay at around 14 degrees South.* Image: © Commonwealth of Australia, ACRES, Geoscience Australia

Shelf margin

The outer margin of the continental shelf is distinctive, and its seaward margin descends abruptly into the Queensland Trough, more than 1000 metres deep. By contrast, most of Torres Strait is less than 15 metres deep. The continental shelf on which most of the reef is established is generally 30 to 60 metres deep, only reaching depths up to 140 metres at the southern end of the reef. Reefs on the outer margin show a series of contrasting morphologies. Six types of shelf margins can be identified.

1. To the north, along the eastern margin of Torres Strait, there is a series of detached reefs with isolated islands, such as the Murray Islands.
2. Between 10 and 11 degrees South, the margin is characterised by tide-dominated, deltaic reefs. These are short reefs (about 4 kilometres long)

that are dissected with delta-like structures on their western side and formed as a consequence of the strong tidal currents that race through the passages.

3. South of these, from 11 to 17 degrees South, where the shelf is narrowest (reaching only 24 kilometres wide at Cape Melville, 14 degrees South), is a series of long reefs, called ribbon reefs, that runs parallel to the reef edge. They have spur and groove structures on their eastern side, and sand sheets accumulate in their lee (see Figure 8.9).

4. Offshore from Townsville, the shelf widens and deepens, and the reefs are predominantly submerged and set back from the edge for much of the margin between 17 and 21 degrees South.

5. South of this is a region called the 'Pompey complex', a reef tract up to 10–15 kilometres wide, with a series of dissected reefs through which there are strong tidal flows. Within the Pompey complex there are occasional 'blueholes', which are deep solutional features formed during a lower sea level. Submerged reefs have been recorded down the forereef to depths of 80 metres.

6. At the southern end of the Great Barrier Reef, where it is more than 260 kilometres wide, the shelf edge is generally reef-less. The Swains and Bunker–Capricorn groups of reefs, with numerous reef islands, are set back from the shelf margin.

Despite its immensity, recent geological investigations have demonstrated that the Great Barrier Reef is a relatively young feature. The reefs that we see form a thin veneer over older landscapes, preserving the legacy of past sea-level variations that have punctuated its formation. Recent changes of sea level mean that the modern reefs have been in existence in their present positions for no more than 8000 years.

Dating the coral reefs

The Coral Sea opened up between 58 and 48 million years ago, when the Eastern Highlands were uplifted, as discussed in Chapter 1 (see Figure 1.6). Over the past 50 million years, the offshore Queensland and Marion plateaux have alternated between periods of terrestrial sedimentation and marine carbonates. Most of these surfaces have subsided into water depths far too great for coral growth, except for a few reefs, such as Osprey, Flinders and Lihou reefs, which do reach sea level and maintain growth far to the east of the modern barrier reef. The Great Barrier Reef is younger than these plateaux – drilling indicates successive periods of reef establishment during sea-level highstands that go back half a million years, but on biological grounds it seems likely that coral reefs have occurred along much of this coast for most of the past 25 million years.

There are numerous outcrops of the underlying bedrock, resulting in more than 600 islands, called 'high islands' with a bedrock core to discriminate them from the low-lying islands (sand cays) composed of reef-derived carbonate sediments. The high islands are especially common between 20 and 22 degrees South (Bowen and Broad sounds). The Whitsundays represents the largest group, but there are smaller granitic islands throughout the Great Barrier Reef and Torres Strait

Figure 8.10 High islands in Torres Strait: *a)* Yam Island (Iama) and *b)* Darnley Island (Erub). Note the fish traps constructed part-way across the fringing reef on Darnley. Photos: J. Leon

(see Figure 8.10a). In some cases, volcanism that formed the islands occurred recently; for example, Murray and Darnley islands in Torres Strait are formed of pyroclastic cones and basalt flows that are a few tens of thousands of years old (see Figure 8.10b). In places, it is clear that the low islands are themselves founded on bedrock close to the surface; for example, small outcrops of granite occur on Tern and Redbill islands, to the south, and these outcrops provide the first clue that modern reefs comprise only a thin veneer over older surfaces. There is strong inheritance, particularly reflecting the shape of older reefs.

Subsurface studies, using seismic profiling and coring, indicate that much of the substrate underlying the Great Barrier Reef is derived from erosion of the eastern margin of Australia and was deposited across the broad continental shelf when sea level was lower (as described below). Modern reefs often form a veneer over former Pleistocene reefs, in many cases directly mirroring the Last Interglacial topography. It is clear that when the sea was lower, as during the last glaciation, the continental shelf was quite different, with the former reefs occurring as well-vegetated, remnant limestone hills, relics of the reefs that flourished during the previous interglacial. These lowstands are recorded in drill cores as erosion surfaces or as solutional unconformities between reef limestones. The major rivers cut into this shelf; for example, the Herbert River appears to have adopted a course through the Palm Islands, and the Burdekin River adopted a more northerly course than at present. The former channels are largely infilled, although a few remain incompletely filled as depressions, called 'wonky holes' by local fishermen.

Fringing reefs

Fringing reefs develop close to shore, reducing wave energy and protecting the land behind them from erosion by the sea. Fringing reefs also produce carbonate sediment that is incorporated with terrestrial sediment into the beach sands that form the isolated beaches at the rear of the reef flat. Ningaloo Reef is a particularly extensive and pristine reef, fringing more than 260 kilometres of coast south of North West Cape (22 degrees South) in Western Australia. Lord Howe Island has the southernmost fringing reef in the Pacific Ocean.

There are isolated fringing reefs along much of the predominantly rocky coast of northern Australia, both throughout the Kimberley and on the many offshore islands such as Groote Eylandt, the Wessel Islands, and Bathurst and Melville islands. These reefs range from 50 to 2000 metres wide, averaging 300 metres wide, and as described in Chapter 5, a distinctive beach type associated with these reefs accounts for over 200 beaches along the Western Australia, Northern Territory and Queensland coast. Where there is a large tidal range and turbid waters, reef flats may be covered with mud and backed by mangroves.

Fringing reefs are an important component of the Great Barrier Reef; more than 750 have been identified, as well as at least 200 incipient fringing reefs, shore-attached or nearshore reefs that do not yet reach the sea surface. Particularly well-developed examples of fringing reefs are more common around inshore high islands than they are along the mainland coast. For example, prominent but narrow reef flats have developed on the windward side of islands in the Whitsundays and around Lizard Island, and

Latitude (°S)	Shelf area (km²)	No. of fringing reefs [incipient]	No. of mid-shelf reefs	Dominant mid-shelf type	No. of ribbon reefs	Reef area (km²)	Reef cover (%)
N of 11	5329	28 [0]	101	planar	0	835	15.7
11 - 12	13 738	22 [0]	234	patches	9	2883	21.0
12 - 13	7042	17 [1]	126	submerged	14	2582	36.7
13 – 14	7324	8 [0]	138	submerged	5	1280	17.5
14 – 15	11 300	38 [4]	129	planar	12	1606	14.2
15 – 16	6186	13 [0]	78	planar	19	716	11.6
16 – 17	7518	24 [0]	70	submerged	0	825	11.0
17 – 18	8136	15 [3]	59	crescentic	1	756	9.3
18 – 19	17 015	35 [7]	93	patches	0	1119	6.6
19 – 20	27 566	12 [16]	225	patches	1	2265	8.2
20 – 21	26 878	206 [19]	240	submerged	2	2259	8.4
21 – 22	38 775	52 [80]	465	planar	2	2202	5.7
22 – 23	26 651	2 [76]	87	submerged	1	289	1.1
23 – 24	14 163	73 [0]	31	planar	0	415	2.9
S of 24	6361	0 [7]	4	submerged	0	23	0.4
Total	223 977	545 [213]	2080		66	20 055	9.0

Table 8.1 *Latitudinal variation of reefs on Great Barrier Reef*
Source: after Hopley, Smithers & Parnell, 2007

broad, leeside bayhead fringing reefs have developed on some islands (for example, Fantome and Orpheus islands), whereas reefs are absent from other large high islands such as Hinchinbrook Island. Isolated fringing reefs occur on the mainland north of Cairns. Those at Cape Tribulation (16 degrees South), where steep rainforest-clad hill slopes abut the shore, are especially noteworthy, but mainland fringing reefs are most common south of Cairns.

Although fringing reefs might appear to form at the shoreline and grow outwards, closer examination shows that many initially formed several thousand years ago, and have change little in extent since then. A suitable foundation for reef establishment has been essential. Fringing reefs have established over a range of foundations, including Pleistocene reef limestone, alluvium and aeolianite (calcarenite). Many fringing reefs, both in Torres Strait and along the Queensland mainland, overlie mud banks. Large fossil corals, particularly flat-topped corals (microatolls), found at the landward margins of reef flats that are emergent at low water, are evidence of a time of prolific coral growth on many mainland fringing reefs. Radiocarbon dating indicates that these record an early stage of fringing reef development about 5000 years ago; the reefs have since built seaward but at a decelerating rate, with little evidence of significant extension in the past 2000 years. In many cases, this seems to have been because of limited suitable foundations, so that the reef was unable to extent into the deeper water. Many fringing reefs appear to be surviving rather than actively growing.

Platform reefs

Reefs occur in mid-shelf on the Great Barrier Reef where corals have colonised topographic highs that form suitable foundations. These mid-shelf reefs grow upwards towards sea level, followed by lateral consolidation, and they can be classified in terms of their stage of development, with the largest and most mature forming prominent platform reefs.

Initially vertical reef growth over the higher points of the prior topography accentuates the antecedent relief. Mature phases are evident when the reef grows up to sea level and vertical growth is hindered but sedimentary infill of the interior occurs.

The evolutionary sequence of reef types comprises three juvenile stages (unmodified antecedent platforms, submerged shoals and irregular reef patches), two mature stages (crescentic reefs and lagoonal reefs), and senile planar reefs, as shown in Figure 8.11. Crescentic reefs, shaped like a crescent along the windward margin of the platform, are the most common form across the Great Barrier Reef, particularly between 14 and 19 degrees South.

Lagoonal reefs, which have a near-continuous rim around a central lagoon, occur predominantly south of 19 degrees South, especially in the Pompey and Swain complexes (see Table 8.1). Planar reefs, which have an emergent, largely barren surface, are most common north of 16 degrees South. It is these reefs that most often have islands on them because they have near-horizontal and partially emerged surfaces across which wave refraction patterns focus sediments onto islands.

The evolutionary sequence shown schematically in Figure 8.11 implies that reefs progress from one stage to another as the reef grows and the depressions are infilled with sediment. The rate appears to be highly individual. Some small reefs near the mainland reached the senile phase 5500 years ago; different parts of any one reef can be at different stages of development, and some planar reefs such as Wheeler, Fairfax, and Wreck reefs passed through the lagoonal stage in less than 1000 years.

Reef response to sea level

The modern reefs we see form a veneer over suitable foundations, and in many cases these are older Pleistocene reefs preserved as topographic features. In order to understand how these reefs have formed,

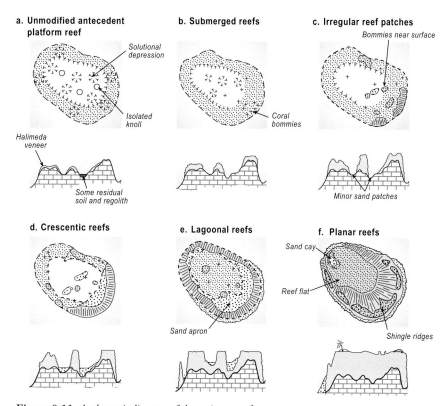

Figure 8.11 *A schematic diagram of the main types of platform reef, in planform and profile, found on the Great Barrier Reef.* Source: after Hopley, Smithers & Parnell, 2007

we look at past changes of sea level and interpret how reef growth has responded to those changes. Fortunately, evidence of past environments is often well preserved in reef limestone, the elevation of which provides an indication of the level of the sea, and a range of radiometric dating techniques enable us to determine their age. During recent decades it has been recognised that sea-level oscillations, as described in Chapter 1, have played an important

Figure 8.12 *Raised reefs of Pleistocene age indicate former sea-level highstands; the series of terraces that marks former reefs to the northeast of Christmas Island in the Indian Ocean has been uplifted by flexure as the island migrates over a bulge in the ocean floor prior to subduction into the Java Trench.* Photo: C.D. Woodroffe

role in reef evolution, and this has been incorporated into the Darwinian model, enabling a clearer interpretation of the response of reefs to those sea-level changes.

The best evidence for past sea levels comes from tropical coasts that have been rapidly uplifted and on which a sequence of emergent fossil reefs is preserved like a staircase. Those on the Huon Peninsula in Papua New Guinea, on the leading edge of the Australian plate, formed the basis for deciphering the pattern of sea-level changes over the past 240 000 years, as portrayed in Figure 1.7.

The Australian continent is not experiencing the rapid uplift rates that are seen at plate margins, so does not have such impressive suites of emergent reef terraces as are found on the Huon Peninsula. However, Christmas Island, an Australian territory in the Indian Ocean, has experienced gradual uplift as the Australian plate migrates over a bulge prior to subduction into the Java Trench. There is a series

of prominent uplifted reefs on this island, formed during successive highstands of the sea. The Lower Terrace, at elevation of around 12 metres above sea level, dates from the Last Interglacial (see Figure 8.12). A fossil reef of similar age is found along parts of the coast of Western Australia, occurring at Fairbridge Bluff on Rottnest Island (32 degrees South), further south than reefs presently form on this coast. This suggests that the Last Interglacial may have been warmer than the present interglacial (usually termed the 'postglacial'). Fossil reefs of Last Interglacial age occur at up to 10 metres above present in the Cape Range National Park, on the central Western Australia coast.

Reefs of the Last Interglacial

Evidence from around the world indicates that sea level during the Last Interglacial was about 6 metres above present. Reefs that formed during the Last Interglacial are found above that elevation where the coast has been uplifted, as on Christmas Island, but below that level where the land has subsided, as on the Cocos (Keeling) Islands (see Figure 8.6), or through lowering of the surface by erosion. The reefs that flourished during the Last Interglacial would have been completely exposed and subject to widespread erosion during the last glaciation. As seen in Figure 1.7, the period of lower sea level persisted for much of the past 100 000 years, reaching a maximum of around 120 metres below present at the peak of the ice age, about 20 000 years ago.

During the ice ages, which have dominated most of the past 2 million years, the sea level has normally been lower than present, and the broad shelf that today supports the Great Barrier Reef has been exposed. For most of this time it would have been a forested coastal plain, with steep limestone hills where there are reefs today; the interglacial periods during which reefs re-established across the shelf were brief in comparison. The Last Interglacial persisted for around 10 000 years, which as we will see is about twice the time that modern reefs have had to develop. Reef organisms must have persisted in isolated refuges, particularly down the steep margin of the outer reef, where there has always been suitable habitat, although adjusting down and back up the reef front, tracking sea level.

Modern reefs

Modern reefs are all younger than 8000 years old, having become established since sea level once again flooded the shelf, following a short period of coral re-establishment over the inundated soils. The pattern of postglacial sea-level rise has been determined from a range of sites around Australia (see Figure 1.7). For example, dating of cores from the Abrolhos Islands off the coast of western Australia shows a trend of rapid rise to around 6000 years ago that is broadly consistent with independent evidence from the dating of shell material in sand barriers from southern Australia and mangrove wood and peat from northern Australia. Modern reefs across northern Australia, like those around much of the western Pacific Ocean, are directly founded on similar Pleistocene reefs, many dated to the Last Interglacial.

Reefs flourished during the final stages of postglacial sea-level rise, as the sea flooded the broad shelves on both continental margins (for example, the Great Barrier Reef) and oceanic reefs (for example, the Cocos (Keeling) Islands). Fringing reefs appear to have established on the mainland coast before 5000 years ago, as indicated by flat-topped fossil microatolls that are exposed during low tide on the inner reef flats, above the low-tide elevation to which their modern counterparts can grow. These provide evidence that the sea level was higher than it is at present, although there remains disagreement about whether this results from increased incidence of stormy weather, warmer sea-surface temperatures, larger waves reaching the coast because the outer reefs had not grown to sea level to reduce wave energy on the mainland coast, or from hydro-isostatic flexing of the continental shelf in response to increased volumes of water in the oceans. After rapid reef growth associated with rising sea level, rates of reef growth have slowed since those reefs have reached sea level, reflecting the limited foundations available for further reef expansion.

In mid-shelf, where it is deeper, the final stages of sea-level rise had been too fast for reefs established on topographic highs to keep pace. These reefs have grown vertically at rates of up to 5 to 8 millimetres per year, and not all have caught up with sea level. Whether or not they have reached sea level, and the extent to which they have infilled, explains the different types of reefs illustrated in Figure 8.11, and the nature of the platform reefs now seen on the Great Barrier Reef. Remarkably, the vertical growth of reefs does not seem to have been slower at higher latitudes, with reefs flourishing on the Abrolhos Islands (28 degrees South), where they accreted vertically at 10 millimetres per year, and on Middleton and Elizabeth reefs, where reefs had caught up with sea level by 5000 years ago, as well as around Lord Howe Island.

Figure 8.13 The pattern of reef growth, and infill of a reef platform, on the Great Barrier Reef, based on coring and dating studies at One Tree Island. Sea level 7000 years ago was below present (SL), but it was above present 5000 and 3000 years ago. Source: after Hopley, Smithers & Parnell, 2007

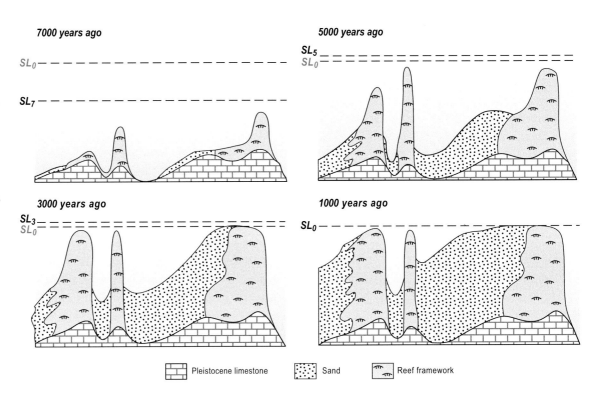

Pattern of reef infill

The pattern of infill of an individual reef is illustrated schematically in Figure 8.13, based on coring and dating of One Tree Island in the Capricorn Group. Vertical reef growth was vigorous between 8000 and 7000 years ago, after the sea flooded the continental shelf. About 80 per cent of the framework of the modern reef was laid down during this period, with massive corals dominating windward, and branching corals dominating leeward reefs. Once sea level had stabilised around 6000 years ago, the lagoon continued to infill through patch-reef growth and sediment accumulation. One Tree Island appears to have been a crescentic reef between 5000 and 4000 years ago, and an open lagoonal reef from 4000 to 3000 years ago. Davies Reef is an example of a lagoonal reef that is infilling with sediment (see Figure 8.14).

Planar reefs

Planar reefs represent the final stage of reef infilling. They have an extensive platform surface that is emergent in cases where there has been a slight fall of relative sea level, which provides a suitable substrate for the development of reef islands. Island sediments record an incomplete history of deposition, and such accumulation seems likely to continue, interrupted

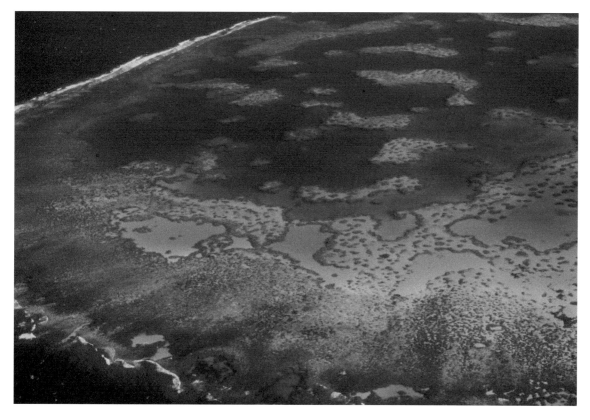

Figure 8.14 *An infilling lagoonal reef, Davies Reef, on the Great Barrier Reef.* Photo: D. Hopley

by erosional events, so long as sediment production continues and sediment transport is possible. In the following section we describe the different types of reef islands that develop on the Great Barrier Reef and in Torres Strait.

Reef islands

In the Great Barrier Reef there are about 300 reef islands, called 'sand cays' – the small, low islands composed of carbonate sediments found on reef platforms, with another 100 or more low islands in Torres Strait. Reef islands are best developed on emergent planar reefs (see Table 8.2), on which the reef flat is composed of massive corals that grew around 4000–5000 years ago, when sea level was higher than it is now. Their form is a function of sediment type, as well as its mobility, location and shape of reef platform, wave conditions and the vegetation that has grown on them.

Reef islands can be composed of sand or shingle, or a combination of both. For example, Lady Musgrave Island is formed of mixed shingle ridges that are often cemented at their base and deposited

	Submerged reefs	Reef patches	Crescentic reefs	Lagoonal reefs	Planar reefs	Ribbon reefs	Fringing reefs
Number of reefs	566	446	254	270	544	66	545
%	21	17	9	10	20	3	20
Mean area (km²)	6.2	9.1	16.8	15.7	4.1	16.4	1.0
Number of unvegetated reef islands	0	42	14	11	135	4	7
Number of vegetated reef islands	0	4	2	0	83	0	0
Number of low wooded islands	0	0	0	0	42	0	2

Table 8.2 The number of reefs of different types on the Great Barrier Reef, and the occurrence of islands on them. Source: after Hopley, Smithers & Parnell, 2007

during successive storms, but the island is encircled by sandy beaches. Unconsolidated sand and shingle can become partially fixed in position as a result of colonisation by vegetation. Vegetation comprises grass and sedge, but on older or larger islands, shrub and woodland indicate greater stability. Woodlands of *Pisonia grandis* used to be common on the larger islands, and these trees attracted numerous seabirds that nested in their branches. Bird droppings (guano) contributed to phosphatic soils on a few islands, giving rise to commercially viable deposits of phosphate, which were mined on several islands. Lady Elliot Island, built from a succession of shingle ridges (see Figure 8.15a), supported *Pisonia* trees and was heavily phosphatised, and almost all phosphate there has been mined. In places, carbonate cements have lithified beach sands into beachrock (see box 'Beachrock', on page 160), which indicates former beach position and implies that the island was stable long enough for sands to become cemented. Radiocarbon dating provides insights into the age of sediment, providing a clue to the time of deposition of the islands and indicating incremental

accumulation over past millennia on both Lady Elliot Island, at the southern end, and Warraber Island (see Figure 8.15b) in Torres Strait, at the northern end of the reef. Although reef islands are composed of sand, they are usually too small and the sand is often too coarse for dunes to develop. Reef island stability is of considerable importance where islands support village communities, as in Torres Strait.

Unvegetated cays

A series of different types of reef islands is illustrated schematically in Figure 8.16. Unvegetated cays are the simplest form of island. These tend to be highly mobile, changing in outline in response to seasonal changes and extreme events such as cyclones. They are most often composed of sand, but exceptionally may be shingle (for example, Pandora Reef). There are numerous unvegetated cays north of Cairns, but they are generally absent south of Cairns, until the Pompey complex and the Bunker–Capricorn groups in the southern part of the Great Barrier Reef.

Vegetated reef islands

Vegetated reef islands form predominantly on planar reefs; 15 per cent of planar reefs have islands on their top but most planar reefs do not, perhaps because waves are too powerful for sediment to accumulate. Vegetated islands are generally sandy (see Figure 8.17), but multiple islands such as Fairfax and Hoskyn islands are examples of where a single platform has a vegetated sand island at one end and a shingle island at the other. Mangrove islands are rare on the Great Barrier Reef (although Murdoch Island is an example), but they do occur in Torres Strait.

Figure 8.15 a) Lady Elliot Island, on the southern Great Barrier Reef, has built towards the southeast (foreground) through accretion of shingle ridges on this planar reef.
Photo: S. Woodroffe;
b) Vegetated sand cays support Torres Strait islander communities on Warraber (Sue) Island in central Torres Strait.
Photo: J. Leon

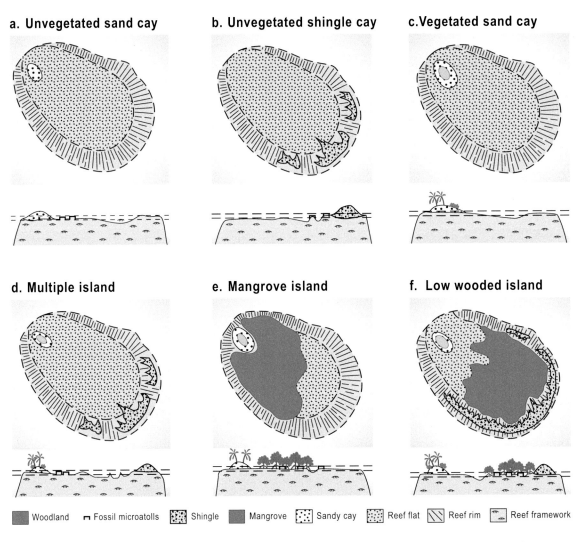

Figure 8.16 A schematic diagram of the main types of reef islands, in planform (above) and in cross-sectional profile (below), found on the Great Barrier Reef. Source: after Hopley, Smithers & Parnell, 2007

a. **Unvegetated sand cay**

b. **Unvegetated shingle cay**

c. **Vegetated sand cay**

d. **Multiple island**

e. **Mangrove island**

f. **Low wooded island**

Woodland   Fossil microatolls   Shingle   Mangrove   Sandy cay   Reef flat   Reef rim   Reef framework

Low wooded islands

A particularly distinctive type of island is called a 'low wooded island' (see Figure 8.16). This type of island occurs on emergent planar reef platforms, and so is most common on the inner shelf. They have shingle ramparts on the windward margin, deposited during cyclones, that are typically tens of metres wide and up to 3 metres high, and a leeward sand cay. The centre of the platform is covered to varying extent by mangroves. In many places, coral microatolls protruding from the reef flat

Figure 8.17 An
example of an
uninhabited vegetated cay
in central Torres Strait.
Note the outcrops of
beachrock along several of
the shores, which give the
island greater stability,
and the dense vegetation.
Photo: J. Leon

are evidence of a former sea level higher than present, though many are obscured by mud accumulating within the mangroves (for example, Houghton and Bewick islands).

Stresses on reefs

Although a coral reef might appear to be a robust geological feature with a persistent rim of jagged coral that withstands the highest waves, in fact it supports a fragile ecosystem that is subject to a range of stresses. These include periodic natural disturbances, such as storms – a record of the impact of storms is recorded in the geomorphology of shingle ramparts, boulder ridges and conglomerate deposits. An incomplete archive of palaeoenvironmental information is contained in the reef limestone and associated sediments and, at a finer scale, within the geochemistry of the skeletons of long-lived corals.

More insidious are the stresses that the reef is subject to as a result of human activities. These stresses include direct impacts such as trawling of the sea floor by commercial fishing, as well as more subtle ecological responses, such as those we infer to have caused infestations of starfish (see box 'Crown of Thorns Starfish', below), and indirect consequences that may result from changes in land use in adjacent catchments and runoff onto the reefs.

Reefs worldwide are under major stresses, from such factors as harmful fishing practices, pollution from coastal development and deteriorating water quality from increased sediment and nutrient inputs. Archaeological evidence from rock shelters, fish traps, middens, hearths and tools composed of exotic rock materials indicate intermittent human use of Australian reef resources. There has been a wider range of activities since European settlement, such as mining of guano from several islands including Raine, Lady Elliot Island and Ashmore Reef in the

CROWN OF THORNS STARFISH

One of the first concerns that reefs, in particular the Great Barrier Reef, might be facing disaster arose in 1970, when there was widespread alarm about infestations of the Crown of Thorns Starfish, *Acanthaster planci*. The Crown of Thorns Starfish exudes its stomach over live coral polyps and devours them, leaving vast tracts of bare coral. The cause of the starfish population explosions remains controversial, and there are three alternative hypotheses: first, that the starfish reached plague proportions due to removal of natural predators, particularly the giant triton that had been commercially harvested; second, that the number of adult starfish outbreaks was a response to nutrient enrichment; and third, that fluctuation in starfish populations is a natural phenomenon.

Outbreaks of Crown of Thorns Starfish in the 1970s caused dire predictions in the popular media that coral reefs as we know them might become extinct. Conservationists mounted a major program to save the Great Barrier Reef, focused on manual removal of starfish, whereas tourism operators and government committees of inquiry adopted different perspectives. Scientific studies have identified fragments of the starfish spines in cores from the Reef, showing that it has been a part of the ecosystem for millennia. However, the cause of the outbreaks is still not understood. Today, the problem remains undiminished, but has been overshadowed by concerns about other threats to the Reef.

Timor Sea, and use for navigation aids (for example, Low Isles), military targets (for example, Rattlesnake and Fairfax islands), collection of Bêche-de-mer (*Holothuroidea*, also known as 'sea cucumbers' and *trepang*, and dried as a food; Lizard Island) and turtle processing (Heron Island).

Changes in land use

Major changes in land use have occurred since the 1870s in many of the catchments that drain into the waters of the Great Barrier Reef. The forests that may have been subject to periodic burning off by Indigenous people were exploited first by logging, then were cleared for grazing, and along the coast intensive sugar-cane cultivation commenced on the coastal plains. Sugar cane now covers more than 4000 square kilometres in Queensland, with a major concentration around the Herbert River. Concern has been raised that these activities, particularly fertiliser application, may have resulted in greater nutrient or sediment loads than can be tolerated by corals.

Delivery of nutrients and sediments

The combined catchment area of the rivers that drain to the Great Barrier Reef is 420 000 square kilometres, roughly twice the area of the shelf beneath the Reef. The two largest, the Burdekin River (130 126 square kilometres) and the Fitzroy River (142 537 square kilometres), account for 64 per cent of the area, and more than 30 smaller and steeper rivers drain the remainder (see Chapter 4). Rainfall is geographically variable, but regularly exceeds 6000 millimetres a year in catchments such as the Tully River, where the Great Escarpment rises to 1000 metres. The total discharge from gauged catchments averages about 40 cubic kilometres per year, which can be extrapolated to around 70 cubic kilometres for all catchments. This might seem like a lot, but it is only about 1 per cent of the water overlying the Great Barrier Reef shelf and less than half the direct rainfall received on the Reef itself.

The rivers today carry large sediment loads, estimated to be 15 to 22 million tonnes of sediment a year. This compares with computer model calculations that indicate the pre-1850 sediment delivery to have been 1 to 5 million tonnes. Discharge and sediment delivery are highly variable, reflecting seasonal variations, long-term patterns such as El Niño Southern Oscillation (ENSO), and the episodic effects of cyclones. Big flood events occurred in 1974 and 1991, whereas the 1930s, 1960s and 1980s were relatively dry decades. A proxy record of sediment delivery to the Reef is preserved in long-lived corals, and the Burdekin River's influence has been reconstructed back for about 350 years, based on fluorescence in the annual bands of corals from the Palm Islands, within the area influenced by the river's sediment plume.

The pre-1850 nutrient delivery is estimated to have been around 23 000 tonnes of nitrogen and 2400 tonnes of phosphorous per year, on average. Modern delivery rates are estimated to be 43 000 tonnes of nitrogen and more than 7000 tonnes of phosphorous, on average. The extent to which these elevated levels are impacting the Reef remains controversial; there are corals that thrive in turbid conditions on shallow nearshore reefs, but elsewhere algae are growing over the surfaces of degraded reefs and are limiting coral recovery. Some terrestrial input of water, sediment and nutrients has occurred naturally in the past, and distinguishing the levels of such inputs that reefs can tolerate is a challenge for sustainable management.

Climate change

Human actions may be having other, more subtle impacts, and recent concern has been raised about the threats to coral reefs posed by climate change. Coral bleaching in response to thermal stress is the principal concern, but additional threats include ocean acidification, sea-level rise and the effects of more intense or frequent cyclones. Coral bleaching occurs when surface waters exceed the seasonal maximum temperature, to which coral and their symbiotic zooxanthellae are adjusted, by 1–2 degrees Celsius. Some recovery may occur, but many corals die if bleaching persists, or temperatures

are particularly high, as was associated with the El Niño year of 1998 in much of the Indo-Pacific and on the Great Barrier Reef in 2002. Global climate models indicate that this temperature threshold will be exceeded more often in the future, with recurrent bleaching at frequencies that reefs cannot sustain; this is discussed in more detail in Chapter 9.

Recently, it has also been suggested that ocean acidification, the slight reduction in pH of the sea that occurs as increased carbon dioxide is taken up into the oceans, is likely to weaken the carbonate skeleton of corals, and thus further threaten reefs.

Sea-level rise may pose additional threats to the islands that occur on the top of reef platforms. The pattern of sea-level change over the past few millennia has varied spatially, but in many cases a reef platform has experienced a slight fall of relative sea-level, and this has assisted islands to accrete on the emergent reef flats. Sea-level rise could have various effects over such reef flats. Increased wave activity may mobilise more sediment, moving it towards existing reef islands and increased wave run-up might build higher ridges on the shore of the islands. Alternatively, increased wave energy across a reef flat could induce shoreline erosion. Islands differ in their resilience, reflecting their geomorphology and vegetation, and islands on which natural processes of erosion and redeposition remain little altered seem most likely to cope with such additional stresses.

Coral reefs have survived the natural climatic and sea-level changes that have occurred over millennia, but the synergistic effects of multiple anthropogenic stresses now pose a far greater threat. The resilience of reefs can be increased by reducing other stresses, such as overfishing and pollution. Good management may lessen the impact or enable the reef to recover, depending upon the extent to which it is already degraded. The Great Barrier Reef, and other near-pristine reefs in northern Australia, may be impacted by aspects of climate change, but they also provide important reference sites, being areas where other human impacts are minimal, and against which the more stressed reefs of the highly populated coasts in the Indo-Pacific region can be compared.

Coral reefs around Australia

Torres Strait

The extensive networks of reefs, together with the treacherous tides, have rendered Torres Strait a navigational hazard for more than 400 years. Warrior Reef, for example, is 34 kilometres long, most of its 165 square kilometres are covered with seagrass and it does not contain any islands. The shallow shelf has been a land bridge between the south coast of New Guinea and the tip of Cape York Peninsula for much of the past, and the islands remain as stepping stones now that the sea has again flooded the shelf. There is a rich archaeological and anthropological record and a vibrant indigenous islander culture. Torres Strait Islanders have established communities on 17 islands, but make wider use of the numerous small and uninhabited islands and the reefs around them. The islands include low alluvial islands with mangroves, low coral cays and a series of high granite and basalt islands, the most important of which are Thursday Island, the administrative centre, and adjacent Horn Island, which has the main airport. Within central Torres Strait there are a number of emergent platform reefs with sand cays, such as Warraber (Sue), Coconut

(Poruma) and Yorke (Masig) islands. In the east, the Murray Islands (Mer) and Darnley (Erub) are young volcanic islands with fringing reefs around much of their perimeter, and with more extensive reef tracts close by (see Figure 8.18).

The Great Barrier Reef

The Great Barrier Reef, extending south from Torres Strait over more than 14 degrees of latitude, is the largest and best-managed reef, and the largest marine protected area, in the world. Since 1981 it has been recognised as a World Heritage Area because of its outstanding natural values, and is jointly administered by the Australian and Queensland governments. It is the focus of a major tourism industry, which attracts nearly 2 million visitors each year, and its natural resources underpin about $7 billion worth of economic activity per annum. Commercial fisheries, employing around 3600 people, have a gross value of about $119 million annually, whereas recreational fishing and boating contribute more than five times this to the region's economy each year.

The rich diversity of reef types and island shorelines has been outlined above, and is a feature of the Great Barrier Reef. There remain many remote and inaccessible reefs, such as Tijou Reef, a ribbon reef to the north, or the dissected and inspiring Pompey group, to the south. To the north, much of the outer ribbon reef is relatively close to the mainland and is a fairly continuous linear reef, broken by occasional passages. The number of mid-shelf reefs increases southwards. Small fringing reefs occur at places on the mainland, such as near Port Douglas, with the transition from rainforest to reef distinguishing this part of north Queensland as a drawcard for

tourists. The southern Capricorn–Bunker group consists of many small cays, such as Heron Island. Each island is different in its particular way: both Magnetic Island, which is a granitic island close to Townsville, and Green Island, a sand cay close to Cairns, receive many visitors each day, whereas there are resorts on more remote islands such as Lizard Island, the Whitsundays and Lady Elliot Island. Tourism focuses on the wondrous reef ecosystems, major seabird and turtle nesting sites on the islands, or a variety of water-based leisure activities. Good management is based on research, and a number of the islands support research facilities. Heron Island has the longest history of research, but research stations have also been established at Low Isles, One Tree Island, Orpheus Island, Lizard Island, North Keppel Island and Horn Island in Torres Strait.

Reefs in the Coral and Tasman seas

Lord Howe Island has a fringing reef along 6 kilometres of its western shoreline (see Figure 8.7a).

Figure 8.18 *Among the Murray Islands in Torres Strait, lies Waer, which is a remnant of a volcanic cone around which there is a fringing reef that has provided sufficient sand for a carbonate beach to accumulate. Mer is seen in the background.*
Photo: J. Leon

Figure 8.19 *This Landsat satellite image of the southernmost Pelsaert Group within the Abrolhos Islands, off the coast of Western Australia, shows the windward reef to the west and the island to the southeast.*
Image: © Commonwealth of Australia, ACRES, Geoscience Australia

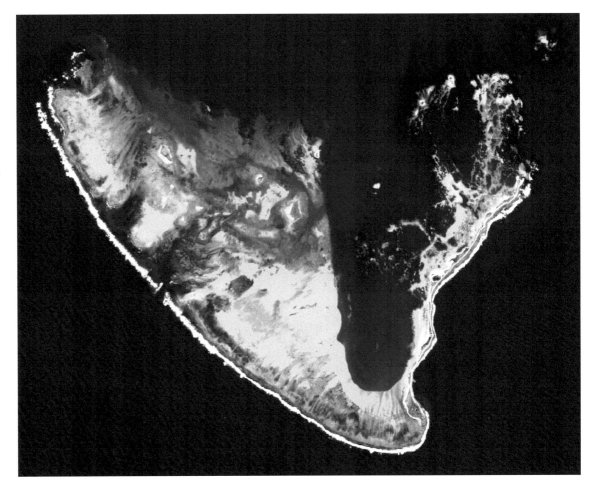

The reef crest encloses a shallow lagoon that is up to 2 kilometres wide and 1.5 metres deep at high tide, with several isolated deeper holes up to 10 metres deep. The coral communities are dominated by branching corals, and although these are at the poleward limit for reefs, coral coverage has increased over recent decades. There are reefs around parts of the coast of Norfolk Island, which posed a hazard for early mariners, but the isolated reefs in the Coral and Tasman seas have been even more sinister. Isolated reefs, such as Lihou Reef, occur in the Coral Sea. Elizabeth (29 degrees 56 minutes South) and Middleton (29 degrees 27 minutes South) reefs are two atoll-like reefs in the Tasman Sea that have had a relatively long and infamous history because of the number of shipwrecks. The rim that surrounds each reef is producing coral, coralline algae and foraminifera (microfossil) sediment, which is progressively infilling the lagoon from the margins. The reticulate mesh of reefs is being smothered by prograding sand sheets. Rarely does the sediment accumulate on the rim, except for minor ephemeral sand cays on the leeward margin, or accumulations against the debris from the wrecks of former ships that have foundered on the reef (see Figure 8.7b).

Abrolhos Islands

Although there are isolated corals at Rottnest Island, they do not presently form reefs on this island as they did during the Last Interglacial. Instead, the southernmost reefs in the Indian Ocean reaching just south of 29 degrees South are at the Abrolhos (sometimes called the Houtman Abrolhos) Islands. These shelf-edge reefs comprise three groups: the southernmost Pelsaert Group (see Figure 8.19), the central Easter Group and the northern Wallabi Group with associated North Island. There are high-energy, swell-dominated reefs along the windward, southwestern margin, subject to waves averaging 2–3 metres high approaching from the south and west, but dominated by fleshy macroalgae.

The extensive leeward reef areas to the east are covered with coral, in places with a conspicuous bluehole topography. Islands formed of Pleistocene limestone occur on the southeastern margin of the reef and are covered by scrubby vegetation typical of this semi-arid climate. The blueholes of the leeward reef reach depths of 20 metres or in some cases 30 metres, with branching coral around their rims, and with steep and irregular coral-covered walls. Turbidity and cold-water stratification limit coral infill of these holes. Although the dissected reef topography, and particularly the presence of blueholes, implies significant erosion of the reef, coring and dating has indicated that these reefs have grown rapidly over past millennia, with more than 20 metres of Holocene reef formed over the Pleistocene basement.

Ningaloo Reef

Ningaloo Reef lies along the western side of Exmouth Peninsula, and the Ningaloo Marine Park stretches for some 260 kilometres, from Bundegi Reef in Exmouth Gulf, around North West Cape to Amherst Point, south of the small resort town of Coral Bay (21–24 degrees South). Ningaloo Reef is the largest and most pristine fringing reef in the world. It is up to 6 kilometres wide, with an average width of 1.7 kilometres. In places the Reef could be considered a barrier reef because the reef crest is separated from the mainland by a lagoon, but for much of its length fringing reef flats extend to the shore. Much of the Reef is straight, but where there are passages there is a significant break in the Reef and a greater diversity of corals (see Figure 8.20).

The arid climate enables prolific coral growth much closer to the shore than would otherwise occur if there were run-off or sediment delivered from a wetter hinterland. The area is exposed to moderate to high Southern Ocean swell and occasional tropical cyclones, and experiences a tidal range of around 1.7 metres. The Holocene reef that has accreted at Ningaloo Reef is generally less than 30 metres thick, in many places with thicknesses of 10–15 metres, and overlies Pleistocene, largely Last

Figure 8.20 *Western Australia's Ningaloo Reef contains an extensive, long and straight fringing reef, as shown here at Fourteen Mile Camp.* Photo: A.D. Short

Interglacial, reef limestone. The passes through the Reef tend to align with valleys carved by streams draining off the backing Cape Range. The annual visit by whale sharks is a major attraction for Australian and international tourists. These huge filter-feeding sharks gather in Ningaloo Marine Park, following mass coral spawning around March or April each year.

Reefs along the coast of northern Australia

Reefs along the northern coast of Australia are not very well known. There are extensive fringing reef

Figure 8.21 *Fringing reefs in northern Australia often occur along rocky outcrops, as in these examples on Melville Island.*
Photo: A.D. Short

systems along much of the Kimberley coast and particularly around the adjoining islands that lie between 15 and 17 degrees South. There are also extensive fringing reefs along the northern coast of the Northern Territory, and around islands including Bathurst and Melville islands, Cape Cockburn, Goulburn and the Wessel islands (see Figure 8.21). Although these small fringing reefs occur predominantly on rocky coasts, there are over 160 small beaches that are fronted by fringing reefs on the Kimberley coast and a further 35 mainland beaches fronted by fringing reef along the northern coast and around Cape Arnhem. Fringing reefs also extend into the western Gulf of Carpentaria, particularly on parts of Groote Eylandt and the Sir Edward Pellew Group.

In the southeastern Gulf of Carpentaria, coral reefs occur on Vanderlin, Mornington and Bentinck islands. A remarkable group of submerged, largely fossil reefs have recently been discovered in the southern Gulf of Carpentaria, covering an area of 80 square kilometres in water depths of around 30 metres; these grew in what would have been turbid conditions but are now largely bare of any living coral. Corals such as *Turbinaria* can tolerate low light levels and are found in northern Australia where the tidal range is large. When mud accumulates at the rear of reef flats, mangrove forests often fringe the shoreline.

Reefs of the Timor and Arafura seas

On the broad continental shelves of the Timor and Arafura seas, a few shelf reefs reach sea level; these include Ashmore Reef and Cartier Islet, the three reef platforms that form Rowley Shoals, Seringapatam Atoll and Scott Reef. Tidal range on the shelf is relatively large, up to 4.5 metres, and this may explain why these reefs differ from atolls that occur in microtidal, mid-ocean settings. The reefs around the margin of Ashmore Reef, for example, have produced enough sediment that the lagoon is largely infilled with sand. The three islands are not on the rim of the reef, however, as they would be on an atoll, but have accreted on the broad sand flats in the centre of the reef.

Rowley Shoals consists of three ellipsoidal reefs on the open shelf. Mermaid Reef, to the north, is 14.5 kilometres long and 7.6 kilometres wide; Clerke Reef is 15.8 kilometres long and 7.6 kilometres wide; and Imperieuse Reef, to the south, is 17.8 kilometres long and 9.5 kilometres wide, and comprises three lagoons and one passage. These reefs rise up from water depths of about 440 metres and their lagoons have a maximum water depth of 20 metres. Each has a prominent spur and groove pattern around the upper 12 metres of the reef front and then drops off into deep water below 30 metres. Scott Reef occurs 400 kilometres northeast of Rowley Shoals in deeper water (700 metres); it consists of North Reef, which is 16.3 kilometres long and 14.4 kilometres wide, with a lagoon 21 metres deep that has two passages, and South Reef that is 27 kilometres wide with several sand cays. Seringapatam Atoll, 25 kilometres to the north, is 9.4 kilometres wide and has

a lagoon that is up to 30 metres deep; it is a shelf reef and not a true atoll.

Conclusion

Australia has the most extensive and pristine coral reefs in the world, including the Great Barrier Reef, the long Ningaloo fringing reef, scattered reefs around the northern coast, and atolls and other types of reef amongst its island territories. Reefs are biological landforms with a close ecological interdependence between process and form, and reef coasts are the long-term outcome of reef growth and sediment production and transportation, over centuries to millennia. Although the broad distribution of reefs is a response to suitable climatic and oceanographic conditions, the form of modern reefs is inherited from older topography, particularly former interglacial reefs, and has been shaped by the pattern of sea-level change since the last ice age.

Australian reefs have evolved as a response to changes in sea level into a variety of forms – submerged, patch, crescentic, lagoonal, planar, ribbon, fringing, barrier and a range of shelf reefs, each reflecting the underlying foundations, but shaped by contemporary processes such as waves, tides, currents and river influence. Once formed, reefs exert two types of influence on the coasts behind them. They reduce wave energy, protecting the shoreline, and they produce sediment that contributes to the landforms that are built. The outer reef crest of the Great Barrier Reef intercepts ocean swell, resulting in lower, fetch-limited waves in its lee. In the case of fringing reefs, waves break at the reef crest, resulting in considerably lower energy at the mainland or island shore.

Carbonate sediment, formed on the reef, is the dominant material in reef environments. Beaches on mid-ocean reefs are entirely calcareous, whereas the proportion of sediments on mainland beaches in the lee of reefs reflects the productivity and proximity of reef habitats in comparison with generally lesser terrestrial contributions. Reefs are limited or absent where rivers do deliver substantial freshwater, sediment or nutrient loads.

Understanding reefs and the coastal environments that characterise them is an essential prerequisite to good management of these iconic landforms in the future. Coral reefs provide an abundance of natural resources and, in the Australian context, are a major focus of domestic and international tourism. However, they can be disrupted by natural events such as cyclones, and through anthropogenic stresses such as pollution and nutrient run-off. Their future is one of change, in some cases natural, in others influenced to varying degrees by human actions. In particular, there is mounting concern that anthropogenic climate change is having a gradual, but wide-ranging and insidious impact on reefs, which we discuss further in the next chapter.

HUMAN IMPACT
ON THE COAST

Introduction

Indigenous Australians have traditionally used coastal resources and would have retreated inland as the sea level rose, until it stabilised 6500 years ago. Since then they have inhabited the present coast, adjusting to the changes that occurred as beaches and dunes were formed and estuaries infilled, utilising their resources. They left little obvious imprint on the coast.

The first contact between Indigenous Australians and non-Indigenous people occurred along the north coast, with Dutch and Spanish voyagers from the early 1600s and seasonal visits from islanders sailing from Indonesia, particularly Macassar (now Udjung Pandang, the provincial capital of South Sulawesi), which probably became regular by around 1700. These were followed by early European exploration, survey and settlement in the 18th and 19th centuries. The past 200 years have seen the establishment of coastal towns and the growth of major coastal cities and their ports, with estuarine locations remaining important for navigation, transport and trade.

In the 20th and 21st centuries, the Australian population has become increasingly focused on the coast. Over the past 50 years there has been accelerated migration to the coast, with residential, holiday and tourist development in major coastal cities and regional coastal centres reflecting a lifestyle change encapsulated by the term 'sea change'. Coastal development up until the 1970s was focused on managing nature: sea walls and shore-normal structures called groynes were built on some beaches, raw sewage was pumped out to sea and, in places, inappropriate development was approved either too close to the shore or in sensitive wetlands.

A recent change in emphasis has occurred, from reactive 'human versus nature' approaches, as in the case of coastal protection, to the proactive management of dynamic coastal systems. The development of national and state/territory coastal policies and new approaches to coastal management have brought a major shift in the way that we view and develop the coast, in parallel with recent increased concern about coastal and marine environments. The principle of integrated coastal zone management has emerged. This involves comprehensive assessment, and the setting of objectives, planning and management of coastal systems and resources, taking into account traditional, cultural and historical perspectives, and conflicting interests and uses. There is now widespread recognition that the coast itself is an extremely valuable natural resource that must be managed and maintained, and not degraded by inappropriate development. This move towards proactive, collaborative coastal management and policy development has occurred at national and state/territory levels in Australia since the 1990s.

In this chapter, we explore the impact of humans on the Australian coastline over thousands of years, including the use of coastal resources for food, habitat, protection, income and recreation, and attempts to manage the coastal environment for future sustainability. Coastal management and planning goals must take into account the natural processes and hazards – such as cyclones, tsunami and coastal flooding and erosion – that place people, property and the coastline itself at risk. Another

significant challenge is the potential impact of future changes in climate and sea level that threaten our coastal systems. We conclude with a discussion of the prospects for the Australian coast, and the potential to preserve the coastal zone for future generations.

Indigenous Australians

How and when humans reached Australia remains contentious. The area of the Indonesian archipelago between the Sunda shelf and the Sahul shelf represents a distinct biogeographical province (where Asian and Australian fauna and flora have only partially intermixed), identified by Alfred Russell Wallace and known as 'Wallacea' (see Figure 9.1). Although Australia and New Guinea would have been connected by a land bridge during much of the Pleistocene, forming Greater Australia, the islands of Wallacea are in deeper water and would have remained in open sea. Modern human migrants must have island-hopped across the Indonesian archipelago, making sea crossings where necessary. It is speculated that they may have used bamboo rafts, and the individual sea crossings necessary are likely to have taken 1 to 2 weeks.

It has been hypothesised that Indigenous Australians might have arrived as long ago as the Last Interglacial (more than 100 000 years ago), based largely on pollen evidence for vegetation change by intensified burning. However, recent luminescence dating techniques, particularly when applied to rock shelters in western Arnhem Land, imply that 50 000 to 60 000 years ago is a more likely time for these first arrivals, when sea level was lower and sea crossings shorter.

If, as seems likely, the early settlers were coastal people, archaeological evidence is likely now to be submerged on the vast continental shelves that were dry land when they arrived and subsequently have been flooded by postglacial sea-level rise. Both Bass Strait and Torres Strait were dry, making it possible to walk from New Guinea to Tasmania. The Gulf of Carpentaria was also dry, except for a large fresh-water lake, Lake Carpentaria, in its centre. Sea level rose rapidly during the postglacial (see Figure 1.7, page 11), causing the shoreline to migrate across the low-gradient Timor and Arafura shelves. Intertidal communities such as mangrove forests would have had to shift, with likely consequences for the people who used resources from these productive systems. The period of postglacial sea-level rise was therefore one of inland retreat for the coastal communities, as well as increased isolation of the Australian mainland from New Guinea, Tasmania and all the offshore islands.

Early use of coastal resources

Our knowledge of early Indigenous Australians' use of coastal resources comes from several sources. Abundant middens located around much of the coast, many now exposed in estuaries or in the face of eroding dunes, are particularly insightful. Middens consist of long-term accumulations of discarded shells and the bones of animals. They provide a clear indication of sites of early occupation and the types of food collected and eaten. Enormous shell mounds are found around Weipa on western Cape York Peninsula, and substantial middens are preserved at Dunwich on North Stradbroke Island, Dark Point in the Myall Lakes National Park and

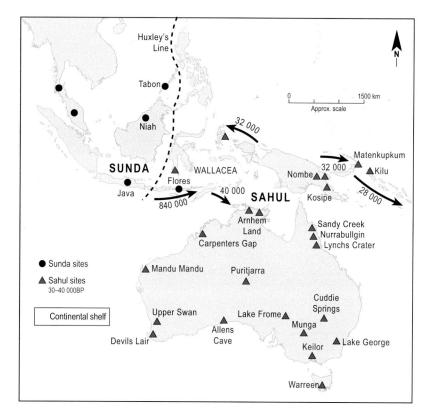

Murramarang Aboriginal Area on the New South Wales south coast. The abundant shells in these can be radiocarbon dated to provide a chronology of the period of occupation (see Figure 9.2).

Rock shelters in northern Australia also contain the remains of shellfish, and provide a record of rock art, together with artefacts such as flakes, grindstones and ochre, that appears to extend beyond the limit of radiocarbon dating, around 40 000 years ago. The portrayal of fish associated with the estuaries, such as barramundi and mullet, as well as crocodiles, is interpreted as marking the onset of estuarine conditions in the Top End of

Figure 9.1 Colonisation of Australia, showing the oldest archaeological sites in Australia and sites of human fossils in Indonesia. Source: after Morwood, 2002

Figure 9.2 *Circular mounds containing shells, particularly the cockle (Anadara), on a rocky knoll overlooking saltflats and mangroves at Darwin Harbour.* Photo: P. Bourke

Figure 9.3 *Rock art, X-ray style paintings of large fish, presumably barramundi, from the Obiri rock shelter in Kakadu National Park.* Photo: C.D. Woodroffe

the Northern Territory (the 'big swamp' described in Chapter 4). It can be inferred that these were painted around 6000–7000 years ago as extensive mangrove wetlands were developed, coinciding with stabilisation of sea level around its present level.

A Freshwater phase followed as freshwater wetlands became progressively established over the plains surface, with paintings of freshwater turtles and distinctive waterbirds such as jabiru and magpie geese. These environmental changes appear to have resulted in expansion of population numbers in much of northern Australia. Estuaries, such as those of the Alligator rivers, provided habitats for bountiful fish, whereas the freshwater wetlands also supported a diversity of bush tucker. This phase was followed by a Contact phase, during which ships, guns and other European items appear in the art, recording successive contact with Macassans, Europeans and Chinese miners and buffalo hunters. Although the first indigenous inhabitants collected

the abundant resources that the coast provided, they left little evidence of impacts, beyond the scraps that they discarded in the middens and their now-fading artwork (see Figure 9.3).

Pre-European contact

The appearance of the dingo, which is a placental mammal descended from Asian dogs, around 4000 years ago is evidence of contact from Asia. Macassans visited the north coast of Arnhem Land (which they called Kaju Djawa) and the Kimberley (Marege). They came for the summer in boats called *praus*, using the northwest monsoon to get here, and returning in May on the southeast trade winds. Their incentive was to collect *trepang*, and the sites at which they camped are marked by lines of stone fireplaces and the occasional tamarind tree. Matthew Flinders encountered a Macassan fleet in 1803, off Arnhem Land, and established that these fishermen had been visiting the northern coast of Australia since at least 1780. Trade was also maintained in canoe hulls, weapons, ochre and fish hooks across Torres Strait. These seafarers left little imprint of interaction with the indigenous population, except for these cultural reminders of their visits.

Early European exploration and settlement

It is convenient to recognise four phases in terms of early European contact and the ensuing settlement and exploration of the Australian coastline (see Figure 9.4). The earliest contact with Australia, of which there is firm evidence, was along the north coast in 1606, when the Dutch landed on western Cape York Peninsula and the Spanish sailed through Torres Strait. The Portuguese were already established in the Spice Islands, and there remains some speculation that they might have visited Australia in the 16th century. During the second phase, beginning in 1616, the coast of Western Australia and parts of the south coast, including Tasmania, were the points of early Dutch contact, although they rarely landed on what they regarded as largely inhospitable shores. The third phase involved Captain James Cook's survey of the east coast of Australia in 1770, and heralded the first European settlements in the late 18th century. This phase, with the establishment of settlements, marked an increase in the impact that people had, particularly along much of the east coast, although

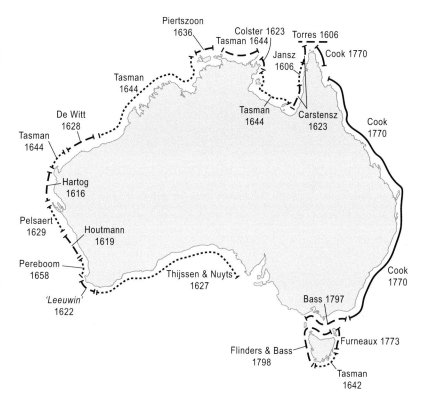

northern Queensland remained largely inaccessible until the fourth phase in the 19th century, because of the extensive reef systems.

The first major coastal settlements (Sydney, Hobart, Brisbane, Melbourne, Adelaide and Perth, see Figure 9.5) were all located on estuaries. They were highly dependent on the sea for transportation, and have maintained their significance as ports to this day. The earliest European settlements in Australia were also penal colonies. These settlers faced severe challenges adapting to the Australian climate and sustaining food supplies. At the same time that settlements were striving to become sustainable, and ships from England were bringing more settlers,

Figure 9.4 European exploration of the coast of Australia during the 17th and 18th centuries Source: after Pearson, 2006

251

Figure 9.5 *All of Australia's state capitals and major coastal cities and towns are located on estuaries, including **a)** Sydney on Port Jackson and **b)** Hobart on the Derwent River.* Photos: A.D. Short

exploitation of coastal resources had begun around the coast. Several hundred whaling ships were in Australian waters from 1798 through the first half of the 19th century. Seals were also hunted until 1810, by which time the seal populations were in decline. Many of these early impacts have not been recorded, although they were likely to have been detrimental.

European settlers exploited coastal lands for their resources. The early history of much of the east and southeast coast is one of initial exploitation of the hinterland for timber. The rainforests adjacent to the coast provided fine timber, in particular Red Cedar (*Toona australis*), and extensive areas were logged around Sydney, with cedar first exported from the Shoalhaven River in 1811. The timber gatherers were closely followed by small fishing communities, and then farmers looking for rich alluvial farmland on the coastal plains. Most of the fertile east coast valleys were converted to intensive agriculture (such as for sugar cane, as described in Chapter 4) by the end of the 19th century. Once farms were established, coastal communities, small towns and ports rapidly followed. The towns usually were located far enough upstream to be safe from flooding but still accessible by ship.

On the New South Wales coast, shipping remained the principal means of transport until the construction of a railway in the 1920s. The first breakwater was commenced around 1820 at Newcastle on the Hunter River, linking a nearshore island to the mainland to improve the port.

Maintaining safe navigable access proved a major challenge at the mouths of many of the rivers in northern New South Wales, because the strong wave conditions produced sand bars that tended to shallow, or even close, the river mouths. This necessitated the construction of jetties, called training walls, the earliest being commenced at Ballina in the 1890s. These training walls were designed to keep the mouth open through the scouring action of river discharge and the ebb tide, and still dominate the entrances to major rivers or estuaries (see Chapter 4). Although improving the navigability of major river mouths, the construction of training walls altered or interrupted the movement of sediment, which in some cases led to significant erosion or sediment build up (as discussed below).

Coastal shipping

From the earliest days of European settlement until the 1950s, much of Australian intrastate and interstate trade, and all of its international trade, was by ship – initially sailing ships and then steamers. This trade supported both large ports and numerous smaller ports right around the coast. Some serviced coastal towns with passengers and freight, while others were used for a single product, such as wool, wheat, coal and timber. The Murray River was used for transport, but the shoals at its mouth hampered shipping, and so a rail link was established in 1853 from Goolwa on the lower reaches of the river, to a jetty built at Port Elliot, although this remained operational for coastal shipping for only a few years. While most of the small ports remain, they are now used by fishing boats and recreational boaters, with the great coast shipping network a thing of the past.

Mining

The growth of mining industries resulted in further need for ports. Discovery of gold in Western Australia in the 1890s led to the construction of port facilities at Fremantle, in the mouth of the Swan River.

The location of mineral resources close to the coast led to the development of mines, ports and, in some cases, associated industries. In New South Wales, coal mining commenced at Newcastle in 1801, using convict labour, and in the Illawarra district in the 1850s, both leading to their development as ports and subsequently as sites for steel making. A port was established in the Illawarra district in 1883, to export coal from the Mount Kembla mine; this became Port Kembla, which underwent a major expansion with the opening of the inner harbour in 1960. Coal mining remains an important extractive industry along the hinterland of the east coast, with coal exports from the Hunter Valley, Illawarra district and central Queensland.

The small port of Robe in South Australia was developed from 1847 as a consequence of the gold rush in Victoria. Whyalla, in South Australia, was established in 1889 to exploit the nearby Iron Knob iron ore resources, which also led to steel making. Hopetoun jetty and railway in Western Australia were established in 1909 to service the copper mine at Ravensthorpe. Bauxite deposits exposed in coastal bluffs in the Gulf of Carpentaria led to the development of Weipa (see Figure 9.6a), Gove and Groote Eylandt bauxite mines, during the 1960–70s. Vast iron-ore deposits discovered in the Pilbara in 1962 resulted in the construction of the ore export ports of Port Hedland, Dampier and Cape Lambert during the late 1960s, and Cape Preston

Figure 9.6 *Exploitation of the coastal zone includes a diverse range of activities, such as **a)** bauxite mining at Weipa, Queensland and **b)** salt farming at Useless Loop, Western Australia.* Photos: A.D. Short

in 2006. Also in the northwest, the development of the northwest shelf gas fields in the 1990s has led to onshore facilities being established at Karratha. Gypsum and salt were also extracted, particularly around Dampier and in Shark Bay (see Figure 9.6b), and salt is also mined and exported from Point Culver in Western Australia. Oil production commenced off the Gippsland coast in Victoria in 1965 and is also piped ashore to Woodside on Ninety Mile Beach. Heavy metal pollution of the coastline has often been associated with mining, particularly in the upper Spencer Gulf, such as at Port Pirie, Port Augusta and Whyalla. There have been significant accumulations of mine tailings associated with some mining operations, such as in Tasmania, where tailings from alluvial tin pollutes the Ringarooma River and material derived from open-pit mining and soil erosion near Queenstown pollutes the King River, which flows into Macquarie Harbour.

Sand mining

Sand represents one of the most important resources on the coast. Quartz-rich beach sand is used as aggregate in construction, to make cement, and is mined in large quantities. Pure quartz sand is mined in lesser quantities for glass manufacturing. Sand is also mined for the rarer heavy minerals, such as rutile, ilmenite and zircon that are used in a range of industrial applications. Sand aggregate is needed in large quantities for road building and housing construction, and in the past beaches and dunes

Figure 9.7 a) Heavy mineral sand mining of coastal dunes is still active in Queensland, on North Stradbroke Island, but has ceased in New South Wales, where it was mined until the 1980s at numerous locations along the coast, including the Myall Lakes, shown in b). Photos: A.D. Short

were seen as a readily available source. Many Sydney beaches were mined for sand, including Narrabeen and Avalon, and sand mining continues at Sydney's Kurnell Peninsula, where the largest dunes in the region have now been removed. Sydney alone requires several million cubic metres of sand a year and, once mining ceases at Kurnell, sources are likely to be expanded onshore or possibly offshore, where massive sand reserves lie on the inner continental shelf.

Silica sand mining occurred in the past at Moore Park in Sydney, to supply the adjacent glass factories. This sand has been blown north from Botany Bay during a period of lower sea level. Silica sand mining continues on Cape York Peninsula, at Shellburne Bay, with the sand exported overseas. Heavy mineral sand mining commenced in the 1930s along the northern New South Wales and southeast Queensland coast, as well as from sand islands, and rapidly expanded into a major coastal industry, although it is now restricted to one mine on North Stradbroke Island. In this case, only the heavy minerals are extracted, usually less than 5 per cent by volume, with the remaining sand redeposited on site and revegetated. Major heavy mineral mining also commenced in the Capel region of southwest Western Australia in 1956, mainly on inner Pleistocene barriers, and minerals continue to be mined and exported via Bunbury.

Extractive sand mining has had the biggest impact along parts of the coast, with the wholesale removal of coastal dune systems, as has happened at Kurnell and Seven Mile Beach in New South Wales, where mining pits remain. Silica mining also removes the dunes but is far less extensive, and while heavy mineral mining disturbs the dunes they are replaced and rehabilitated, with little obvious long-term impact.

Commercial fishing

Perhaps the most obvious coastal resource industries are fisheries, with commercial fishing operating from scores of ports located all around the coast and employing over 20 000 people. Commercial fishing targets more than 200 species of fish, 60 crustacean and 30 mollusc species, with a total catch of over 200 000 tonnes each year. As an example, commercial fisheries employ around 3600 people across the Great Barrier Reef region, with a gross value of around $119 million annually. By comparison, recreational

Figure 9.8 Oyster racks *on the intertidal sand flats at Adams Bay, on Bruny Island, Tasmania.* Photo: A.D. Short

fishing and boating contribute $640 million to the region per year. Coastal ecosystems are particularly important, therefore, in ensuring the sustainability of fisheries. The coast plays a major role in the life cycle of barramundi and prawns, which support large commercial fisheries in northern Australia. Torres Strait supports lobster fisheries. Abalone catches are collected from shallow, temperate marine habitats, and there are numerous more specialised resources, such as mud crabs within estuarine areas.

Aquaculture (called 'mariculture' when in the sea) involves the farming of aquatic organisms and includes a range of ventures, from pearl oysters in the northwest to rock oysters in southern Australia (see Figure 9.8). Salmon and Bluefin Tuna are increasingly raised in aquaculture farms in southern Australia, while prawns are farmed in shallow ponds on coastal lowlands. All these ventures can have a number of impacts, including conflicts with other users of the particular area, loss of visual amenity and deterioration of water quality.

Urbanisation

In 1901, Australia had a population of around 3.8 million. Around one third lived in capital cities, with about half a million in each of Sydney and Melbourne. A century later, the total population had increased by 16 million, and over 85 per cent lived within 50 kilometres of the coast. The capital cities continue to dominate, but 20 per cent of the population now live in coastal towns and cities other than capital cities. In 2006, the Australian population reached 20.7 million, with 63 per cent living in the capital cities. Today, the population of Sydney exceeds 4 million, while that of Melbourne is roughly equal to the entire population of the nation at federation in 1901. Brisbane's population is about 1.8 million, Perth 1.5 million and Adelaide a little over 1 million.

The growth of the major coastal population centres has had a number of impacts on the environment, beginning with massive and ongoing land clearing and replacement of vegetation and soils with buildings and roads. This results in accelerated run-off and stormwater carrying sediment and other pollutants to the coast (see Table 9.1). In cities like Sydney, many of the smaller estuaries and most of the harbour tributaries were reclaimed for ports, industry, recreation and housing, replacing mangroves, salt marshes and seagrasses, and the marine communities they supported. Construction along the open coast and on beaches during the 20th century led to development in hazardous locations and resulted in destruction of houses by storm events. Following this costly and ongoing sea defences have been installed, along with sand nourishment projects, such as along the Adelaide beaches and Gold Coast (see Figure 9.9).

Coastal pollution

Disposal of urban sewage is a major issue in all towns and cities. In Sydney, this led to the construction of deep ocean outfalls to dispose of secondary treated sewage 5 kilometres out to sea. Sewage is processed to various levels; primary treatment involves screening to remove solids and secondary treatment involves removing gross pollutants and organic matter by bacterial action, while tertiary treatment involves further filtering, nutrient stripping and disinfection. In Melbourne, sewage is treated to a tertiary level at the large Werribee sewer farms

Pollutant	Examples	Impacts
Nutrients	nitrogen, phosphorus, iron	Secondary treatment of sewage does not remove these nutrients. Can cause algal blooms
Pathogens	enteric viruses, hepatitis A and E	Bacteria from human and animal wastes found in sewage and stormwater
Toxic organic chemicals	chlorinated pesticides, polynuclear aromatic hydrocarbons	Bio-accumulation in contaminated sediment
Heavy metals	cadmium, arsenic, copper, lead	Bio-accumulation; can be a cause for great concern where metals accumulate in large concentrations
Sediment	soil erosion	Turbidity considered a major deterioration of water quality
Other hazards	oil spills, other wastes	Hazardous to birds and other fauna
Plastics	beach litter, plastics	Hazardous to wildlife and beach users

Table 9.1 Pollutants discharged to the coastal zone Source: after Harvey and Caton, 2002

before being discharged into Port Phillip Bay, while in Adelaide the discharge of tertiary treated sewage into the gulf has been blamed for the die-back of seagrass and resulting shoreline erosion.

On a smaller scale, the problems of land clearing, run-off, pollution and sewage disposal are a concern in every coastal community, with most located adjacent to estuaries that provide important habitat, recreational and fishing venues. Pollution has closed some estuaries to fishing and oyster farming (for example, in New South Wales, a major contamination closed oyster farms in Wallis Lake in the 1990s and commercial fishing was suspended from Sydney Harbour in 2007), while water quality, flooding and sedimentation are a concern in many locations.

In eastern Australia there was a phase in the 1970s during which canal estates were built on

257

Figure 9.9 a) Glenelg, located at the mouth of the Patawalonga River, is Adelaide's most heavily developed and most popular beach suburb. Photo: S. Daw *b) Queensland's Gold Coast is the most densely developed strip of coast in Australia and the site of Australia's biggest tourist strip.* Photo: A.D. Short

low-lying land along the coast, enabling many of the residents to enjoy their own waterfront. However, there are often environmental concerns about these artificial waterways. For example, many were constructed on low-lying coastal plains in areas of potential acid sulphate soils, where formerly waterlogged sediments oxidise to release sulphuric acid (see Chapter 4). Consequently, state governments have largely curtailed the building of such canal estates in New South Wales and Queensland, although they continue to be built in other states.

Since the 1970s, coastal towns have expanded as a result of the 'sea change' phenomenon, with people, young and old, seeking alternatives to living in congested cities. This expansion has been accelerated over past decades, partly by the push of higher costs of living in major metropolitan areas and partly by the pull of alternative lifestyles. There has been growth in coastal cities, those urban centres with populations greater than 100 000, such as Cairns and the Gold Coast, and an extension of the coastal, commuter satellite communities located at the edges of the capital cities, such as Gosford and Illawarra on either side of Sydney, and Rockingham and Mandurah, south of Perth. Whereas average population growth rates of around 2 per cent characterise most communities, some coastal local government areas have seen growth rates in excess of 4 per cent, such as Wanneroo and Mandurah in Western Australia, and Caloundra and Pine Rivers in Queensland. Rapid growth has also typified coastal getaways, such as Bunbury and Busselton in Western Australia and Victor Harbor in South Australia, coastal lifestyle destinations such as Coffs Harbour and Byron Bay and coastal hamlets, such as Robe in South Australia and Port Douglas in Queensland.

Those values that attract people to the coast – the unspoilt scenery and the diverse ecosystems – are often the first casualties of the pernicious spread of urban sprawl. More than 60 local governments from those areas that have already seen population growth have formed a Sea Change Task Force to share information and experiences in the face of this growth. Other areas, such as the central Western Australian coast, covering the 250 kilometres north of Perth, have developed a regional strategy in an attempt to balance urban expansion with conservation, recreation and tourism goals.

Recreational use of the coast

European settlers initially found little use for beaches, apart from being able to launch small boats in more sheltered locations. Sealers, whalers and, later, fishermen occupied those anchorages suitable for small craft within a few decades of settlement. To the general community, beaches were infertile and worthless; the waves were a nuisance and a hazard to all coastal sailors and the dunes behind the beaches a waterless, barren wasteland. Since they were not wanted, or were not taken up as land grants, most Australian beaches and coastal dunes remain in Crown ownership. In the more populated coastal settlements of Newcastle, and later Sydney as it spread out to the coast, the coal miners and workers began to find the beach a source of relaxation on their Sundays off, although the law up to the 20th century forbade bathing between 6.00 am and 8.00 pm, as well as 'mixed bathing'. Despite attempts by the Newcastle coal miners back in the 1860s, it

LEGAL SEA BATHING

At the turn of the 20th century, Australia's outdated bathing laws were confronted by William Gocher, a Manly resident and editor of *The Manly and North Sydney News*. Defying the authorities, he bathed publicly in the surf at Manly Beach, in broad daylight, on a Sunday in September 1902. Gocher did this without incident the next Sunday, then again the next. He was arrested on the third Sunday, bailed out by a friend, and appeared by request before the inspector general of police. In the light-hearted court proceedings that followed, Gocher was acquitted, but the inspector determined that both men and women would have to wear neck-to-knee costumes if they were to bathe in daylight. The thousands who had come across on the Manly ferries to watch Gocher then began to join him on the beach and in the surf. This signalled the official start of daylight bathing, and Australians' great attraction to the beach and surf.

was at Manly in Sydney in 1902 that the law was finally shown for what it was worth, an unpopular anachronism.

Bathing and rescue

The popular beaches of the early 1900s were all either within walking distance of urban centres and towns, such as in Newcastle and Wollongong, or serviced by public transport to the smaller residential settlements, as in Sydney. Bathers soon found, however, that these beaches were as hazardous as they were attractive, and surf lifesaving clubs began to be formed to patrol the beaches. As described in Chapter 5, the southern coast between Fraser Island and Exmouth is dominated by generally moderate-to-high swell and rip-dominated beaches. As a consequence, the increase in beach usage was accompanied by a rash of drownings and rescues, with the establishment of many surf clubs often following a drowning at the beach. By 1914, 34 surf lifesaving clubs were formed in New South Wales,

three each in Queensland and Western Australia, and two in Victoria. The pattern had been set, and a rapid spread of surf bathing in populated coastal areas followed the establishment of surf lifesaving clubs around the coast. The Surf Life Saving Association of Australia (now known as Surf Life Saving Australia) celebrated 100 years in 2007, with 300 clubs now located across the country, together with another 157 beaches patrolled by professional lifeguards. These 457 beaches represent the most popular of Australia's more than 10 000 beaches.

In parallel with the use of the coast for bathing and swimming has been the growth of board surfing, particularly since 1956. Today, more than 300 000 Australians consider themselves to be surfers, with most surfing taking place around the more wave-exposed southern half of the continent.

Tourism

Domestic tourism in Australia has also expanded to make use of the coast, with many rural, urban and city dwellers taking their holidays on the coast. Initially, camping grounds and caravan parks were established along the coast. Over time some of these have themselves become urbanised, and over recent decades there has been substantial growth in these areas in international tourism, which is now the largest international earner for the nation. There is a large tourism industry in Queensland, focused on the Gold Coast and the Sunshine Coast. Large numbers of tourists visit the Great Barrier Reef each year, and this reef-focused tourism is estimated to contribute over $6 billion annually to the national economy. The Great Barrier Reef Marine Park is the world's largest marine protected area and has been managed for multiple uses by the Great Barrier Reef Marine Park Authority since 1975. Coastal venues are increasingly valued for their scenic and environmental qualities, and this has led to the development of eco-resorts and greater tourist participation in water-based coastal activities such as snorkelling, scuba diving and whale watching.

In many cases, it is the coastal ecosystems that attract visitors, and some tourist operators have had to go to considerable lengths to manage the visitors and ensure minimal disruption to the animals that they come to see. Some examples are the Phillip Island Penguin Reserve in Victoria, the Seal Bay Conservation Park on Kangaroo Island, and the Hamelin Pool stromatolites and Monkey Mia dolphins in Shark Bay.

Other users of the coast

Apart from bathing, surfing and tourism, one of the biggest users of the coast is the recreational fishing industry. Recreational fishing is conducted from rocks, the beach, jetties and more than 500 000 boats. More than 3.5 million Australians fish, and over $2 billion are spent each year on fishing, with 80 per cent of fishing taking place on the coast or adjacent seas. Just as controls on commercial fishing have been enacted through quotas, zoning within marine parks and closure of sensitive estuaries, so too have measures such as licences, bag limits and size restrictions been introduced to try to ensure that these recreational fishing efforts do not result in impacts that exceed sustainable levels.

Table 9.2 provides some insight into the extent of usage of the Australian coast. Only 16 per cent of the 11 753 beaches are accessible by sealed road,

	Qld	NSW	Vic	Tas	SA	WA	Kimberley	NT	Cape York	Total	%
SLSC	57	130	53	11	19	25	1	3	0	**299**	
Lifeguard	67	44	18	14	4	9	1	0	0	**157**	
Sealed road	262	402	364	334	230	296	5	28	11	**1932**	16
Gravel road	79	100	40	234	513	283	26	141	20	**1436**	12
4WD	285	100	61	516	694	857	90	253	144	**3368**	29
No vehicle access	575	161	127	533	235	615	1239	1066	466	**5017**	43
	1201	763	592	1617	1672	2051	1360	1488	641	11753	

Table 9.2 Number of patrolled beaches and type of access to all beaches

with a further 12 per cent having gravel road access, meaning you can only drive the family car to 27 per cent of the country's beaches. A further 29 per cent are accessible by off-road vehicle, many over rough and/or sandy tracks, while 43 per cent, or 5000 beaches, have no vehicle access. These are spread around the coast, with the largest concentration across northern Australia. Given that the developed beaches are all accessible by sealed road, and to a lesser extent gravel road, what this tells us is that less than a quarter of the beaches have been developed, with 457 of these being patrolled by lifesavers and lifeguards, while more than three-quarters of the beaches are undeveloped and remain in a relatively natural state. Many of these are located in national parks, reserves and on land owned by Indigenous Australians, so they are likely to stay that way. Another interesting fact is that outside of New South Wales and around the state capitals, most Australian beaches have no formal name. So the typical Australian beach is not only 1.4 kilometres long and bordered by headlands (see Chapter 5), but it is also inaccessible by vehicle and unnamed.

Coastal management and planning in Australia

Coastal management can be defined as the management of human activities and sustainable use of Australia's coastal resources in order to minimise the potential for adverse impacts on the coastal environment, now and in future. Coastal planning, by contrast, involves a longer time frame, and covers the formulation of coastal policies, plans and programs that promote sustainable use of Australia's coastal resources. While these sound like admirable objectives, it is important to realise that such grand aims sit within a political landscape, and have been influenced by a range of external factors. In some cases, Australia can claim that it has been at the forefront of the development of good coastal management practices. It has often been said that the Great Barrier Reef, for example, contains the best-managed reef and coastal waters in the world.

Although there had been previous assessments of the Australian coast and recognition of the need for more effective coastal management, at the national level a House of Representatives (1991) report on *The Injured Coastline*, and the subsequent Resource Assessment Commission (1993) *Coastal Zone Inquiry*, brought a new focus to the key issues. They indicated the inadequacy of management mechanisms, the fragmented nature of decision-making between the three tiers of government (federal, state/territory and local), the lack of community input and a need for longer-term planning. A number of Federal Government initiatives emanated from this, including the development of Coastcare, the Coast and Clean Seas program and a State of the Marine Environment Report.

The 1990s was also a time during which the states focused on coastal policies and their approach to coastal protection and management (state/territory waters extend 3 nautical miles (5 kilometres) from the shore, with the Federal Government responsible beyond that). Although Queensland had enacted the *Beach Protection Act* in 1968, South Australia its *Coast Protection Act* in 1972 and New South Wales its *Coastal Protection Act* in 1979, there was a more concerted effort to take an integrated approach to coastal management in the 1990s. Since then, there has been some further policy development; for example, the Victorian Coastal Strategy has been revised twice (2002 and 2007) and the New South Wales Government undertook a Comprehensive Coastal Assessment in 2006.

The day-to-day responsibility for much coastal infrastructure and planning rests with local government. Councils make planning decisions, approve coastal development, manage beaches and facilities such as boat ramps and amenities, and respond to complaints. Each state/territory government has defined the coastal zone and developed coastal policies and guidelines for coastal management plans, which are then developed and implemented by local government. Increasingly, the states and, to a lesser degree, the Federal Government are co-funding local government coastal management plans and initiatives to enhance the coastal environment. In New South Wales, the state, however, maintains the right of veto on all coastal development.

Protected areas

In the early 1990s, Australian governments identified a need to protect representative marine ecosystems and habitats in marine protected areas, and agreed to establish a comprehensive, adequate and representative system of protected areas covering Australia's exclusive economic zone. A system of Commonwealth Marine Protected Areas (MPAs) is being established to protect habitat deemed to be of high conservation-value, such as individual reserved areas up to 162 000 square kilometres around Macquarie Island, or the recently declared Southeast Commonwealth Marine Reserve network, which comprises 13 reserves in the region around Bass Strait and Tasmania. There is also a range of additional coastal protected areas administered at state/territory level (see Table 9.3). The primary aim of MPAs is to protect a representative sample of biodiversity. Resource exploitation, such as extractive activities, is prohibited where it is considered likely to adversely affect biodiversity, but may be permitted if no impact is envisaged. Similarly, tourism and scientific research are only permitted where they do not harm a fragile ecosystem or an endangered species. Despite the term 'marine', the majority of these encapsulate areas that are coastal.

At the state/territory level, long sections of the coast and coastal waters are also being protected in coastal national parks and marine parks (the marine equivalent of national parks); for example, the Solitary Islands Marine Park, and the recently declared Myall Lakes–Port Stephens and Batemans Bay marine parks and the proposed Twofold Bay Marine Park, all in New South Wales. Many of these

State/territory	Parks & reserves			Indigenous land		Marine parks	
	#	km	%	#	km	#	km²
Queensland	34	1337	22	11	438	3	350000
New South Wales	43	722	45	3	11	6	855
Victoria	13	614	41	0	0	13	531
Tasmania	17	859	28	0	0	7	1235
South Australia	43	1091	29	3	86	19	242
Western Australia	28	2888	28	8	2279	16	18000
Northern Territory	3	832	17	2	2730	1	557
Total	181	8343	27	27	5544	65	371420

Table 9.3 Number and extent of coastal national parks and reserves, coastal land owned by Indigenous communities and area of marine parks, some surrounding offshore islands
Note: The number and area of parks are expanding, particularly marine parks. This table shows the status prior to 2008; % indicates percentage of state or territory and (total) national coastline.

are highly contentious as they impact on commercial and recreational fishing and other water-based activities. However, their aim is to ensure a better managed and more sustainable coastal environment. Australia-wide, 27 per cent of the coastline is presently protected in coastal national parks and state reserves, while a further 18 per cent is land owned by Indigenous communities, primarily in northern Australia, and much of the coast is ringed by marine parks and reserves. These figures are likely to increase into the future, ensuring that a substantial proportion of the coast is preserved for future generations in a natural state. In New South Wales alone, that figure has reached 45 per cent.

More recently, the 2008 Federal Government's 2020 Summit discussed 'Climate change and environmental impacts on coastal communities' as one of its 10 key issues for the future, and this is the focus of an Inquiry by the House of Representatives Standing Committee on Climate Change, Water, Environment and the Arts. It remains to be seen how the results of these initiatives translate into action.

Managing coastal hazards

Coastal hazards place property and the public at risk. They include rare extreme events like tsunami, episodic events like tropical and east-coast cyclones and associated high waves, strong winds and coastal flooding, in addition to more benign processes like long-term shoreline erosion and coastal pollution, all natural events that the coast is able to accommodate. However, where the shoreline has been modified and intruded upon by human settlement the impact can be devastating.

Extreme events

More than 50 tsunami are known to have reached Australia in the past 200 years, with the most vulnerable part of the coast being on the sparsely populated northwest, facing the seismically active Indonesian island subduction zone. The largest event for which a record exists involved a 6-metre-high wave at Cape Leveque in northwest Western Australia in 1977. Geomorphological evidence suggests that large tsunami waves have hit the Western Australia coast in the past; similar evidence along the east coast of Australia has been suggested, but this has been contentiously debated and the occurrence and size of former events remain unclear.

By contrast with tsunami, the impact of tropical cyclones and storm surges is well known. Australia is hit by several severe cyclones each year. The tracks of 25 of the more severe tropical cyclones to hit Australia since 1970 are shown in Figure 2.3 and the devastation caused by Cyclone Tracy is described in the box 'Cyclone Tracy', on page 31. The northern coastline of Australia has a long history with damaging tropical cyclones, and coastal shipping was particularly susceptible to cyclones in the absence of good weather forecasts and adequate communications. Such storms cause enormous wind damage to buildings and extensive flooding as a result of torrential rain, and the accompanying storm surge, which is a result both of low pressure and set-up by winds and waves. Further inundation can result from wave run-up on exposed shores, with extreme water levels reaching a few metres above sea level. Of particular concern is that since the late 1970s there appears to have been a 30-year period of fewer and less-intense tropical cyclones, a period also when considerable development has occurred right along the cyclone belt (see Figure 2.3). If cyclones of the intensity of those seen in the past recur there will be considerable threat to property and people.

Coastal erosion

In most locations, erosion or deposition on the shore are natural processes with usually no impact on human settlement, as a result of which coastal protection of the shoreline is rarely required in Australia. However, in a few locations the dynamic shoreline has become a problem, in some cases a major and expensive problem, and in almost all of these cases the problem is related to human interference or encroachment on the shoreline. Coastal protection works, such as breakwaters, groynes or seawalls, are usually built to guard against erosion. In doing so, they harden the coast and reduce its ability to adjust naturally. A snapshot of several of the erosion hotspots around the Australian coast illustrates some of the issues.

Queensland

Much of the shoreline of the rapidly expanding tropical city of Cairns appears to be especially vulnerable. The popular marina area and the hub of Cairns has encroached on Trinity Inlet, and formerly mangrove-covered mudflats along the seafront are now bare and lined with seawalls, backed by development close to sea level. Much of northern Cairns is built on a coastal plain at the mouth of the Barron River delta. Critical infrastructure, such as the airport, lies close to present high tide level and is threatened by the highest tides, which fill the remaining mangrove-lined tidal creeks that meandered through these plains. Much of the coastal plain and city would be inundated by a 2.5-metre storm surge, calculated to accompany a 1 in 100-year storm. The north Cairns coast along Yorkeys Knob–Machans–Holloway beaches is lined with seawalls because much of the shoreline has been retreating, on a fluctuating coast that in the long-term is building seaward as the Barron River supplies more sediment to the coast. The problems associated with this dynamic shoreline will be exacerbated by climate change.

Australia's best-known strip of defended coastline, and a major holiday destination for local, national and international tourists, is the 35 kilometres between the New South Wales–Queensland border at Point Danger and the Nerang Inlet, better known as the Gold Coast (see Figure 9.10). Here is a system that is part of a conveyor belt of sand moving north from New South Wales, across the border at a rate estimated at 500 000 cubic metres per year. In 1962–64 the Tweed River training walls were extended 400 metres out to sea to keep the mouth navigable. The southern wall blocked the northward movement of sand, trapping millions of cubic metres of sand and preventing it from moving across the border and along the Gold Coast beaches. As the sand supply was depleted, combined with a series of severe cyclones in 1967 (Dinah, Barbara, Dulcie, Elaine and Glenda), 8 million cubic metres of sand was eroded from the beaches and threatened the backing Gold Coast roads, houses and hotels. The solution has been three-fold.

Figure 9.10 The Gold Coast beaches have been maintained by nourishment with sand, pumped from New South Wales via the pier (right) under the Tweed River entrance onto Duranbah Beach (foreground), to continue the longshore drift of sand. Photo: A.D. Short

First, a continuous seawall was built the length of the coast and covered with sand and dunes. Second, between 1995 and 2000, 3.5 million cubic metres of sand was dredged from the Tweed Bar and placed offshore of the southern Gold Coast beaches. Third, a permanent pumping system was built just south of the training wall, which since 2000 has pumped more than 500 000 cubic metres of sand each year from New South Wales, across the border and onto the Gold Coast beaches. In 2008, these beaches were as wide as they have ever been. However, it has all come at a cost in the tens of millions of dollars.

New South Wales coast

A training wall similar to that at the mouth of the Tweed River was built between 1960 and 1966 at the mouth of the Brunswick River, 50 kilometres south of the Gold Coast, to service the small fishing fleet. Studies have shown that the walls impacted the beach to the north and south. The beach built out for 8 kilometres updrift, while erosion was observed to extend up to 17 kilometres downdrift, with the shorelines not stabilising until 1987. The largest impacts were close to the wall, with the small beachfront community of Sheltering Palms, located 2 kilometres north of the wall, experiencing up to 90 metres of shoreline erosion. This resulted in some houses, roads and telegraph poles ending up in the surf zone and, finally, abandonment of the entire village by the mid-1970s. Unlike the Gold Coast, this area did not warrant the massive expenditure on protection, so it was sacrificed.

Coastal erosion, particularly associated with a sequence of east-coast cyclones, has occurred at several points along the New South Wales coast. The 1974 storms, estimated as a 1 in 200-year event, destroyed the pier at Manly and resulted in loss and damage of property, and roads being cut at several points in Sydney and at Moruya, Bermagui and Tathra. Elsewhere, this was the first stage of erosion, with subsequent storms actually undermining property, as with the three houses that were destroyed during a storm in 1978 at Wamberal (see Figure 9.11).

Collaroy Beach, one of Sydney's northern beaches, is a classic example of the inappropriate planning and shoreline subdivision that took place more than 100 years ago. The original property boundaries extended, and still do, down across the dune and onto the beach, with most of the houses and now some high-rise dwellings built on the beach-dune area. The consequences were entirely predictable: every time the beach retreated during high waves, the original beach shacks were undermined. Major erosion occurred in 1920, seven shacks fell into the sea in 1944–45 and one was washed out to sea in 1967. Following the 1945 storms, the council voted to resume the remaining houses. Instead, within 10 years the first block of flats was built and soon after the first high-rise building, which in turn was undermined by the 1967 storms. More high-rise buildings followed, with the next built just in time to be undermined by the 1974 storms. Here, the council had allowed initial development in a hazardous zone, and later massive over-development even after houses had been washed away. Collaroy remains a problem area, with most of the affected properties fronted by makeshift seawalls. The council has started slowly to buy back some properties and hopes the State Government will allow massive beach nourishment at some time in the future. In the meantime, every

Figure 9.11 a) *Following high seas in June 1978, three houses were undermined and fell into the sea at Wamberal (arrow); **b)** One of the destroyed houses.* Photos: A.D. Short

big sea removes the narrow beach and exposes the unsightly and hazardous seawalls on one of Sydney's premier beaches.

The rocky shores of New South Wales appear to be much more resistant against erosion. Locally, however, there are concerns about rock falls and public safety, as at Bilgola, Newport and Narrabeen headlands in Sydney, where remedial works have been carried out. The steep cliffs at Coalcliff, just north of Wollongong, have experienced a series of landslides associated with the claystones and shales that are interbedded with the sandstones. The continual damage to Lawrence Hargrave Drive, and

the threat of falling rocks, led to the construction of the spectacular Seacliff Bridge, a 665-metre stretch of road that is built out from the cliff face over extensive rock platforms.

Victorian coast

Most of the Victorian coast is protected by a foreshore reserve. However, at Portland the reserve was not wide enough to protect a stretch of beach known as the Dutton Way. Here, problems started when a breakwater was completed in 1960 to expand the port of Portland, thereby interrupting the easterly movement of sand to the downdrift Dutton

Way beach (see Figure 9.12a). As the beach began retreating and threatening a road, a seawall was constructed that now winds its way along the shore for 4.5 kilometres.

Low-lying areas around the Gippsland lakes are subject to flooding, particularly after the artificial opening of the Gippsland Lakes in 1889 (see Figure 9.12b). Flooding in 2007, after intense rainfall over the catchments that drain into the lake, was accentuated at high tide and inundated much of Lakes Entrance.

South Australian coast

The Adelaide metropolitan beaches have been experiencing erosion for decades. This is a result of the natural 40 000–50 000 cubic metre per year, northerly sand transport, which is exacerbated by dieback of nearshore seagrass meadows as a consequence of sewage pollution, and further aggravated by some roads and structures located too close to the shore. The erosion has been managed by the construction of 14 kilometres of seawall (see Figure 9.13a), the trucking of sand from the northern end of the system back to the south, and more recently by pumping sand onshore from nearshore sand deposits. Maintenance of these metropolitan beaches continues at significant cost, but it has been possible to sustain the natural values of the coast, even re-establishing dunes in front of the esplanade at Brighton.

Figure 9.12 a) The beach at Dutton Way has been replaced by a 4.5-kilometre-long rock seawall; b) The narrow opening at Lakes Entrance connects the once freshwater Gippsland lakes with the sea. Photos: A.D. Short

In the 1990s, in a political decision that will ensure South Australia has generations of coastal problems, the South Australian government voted to freehold more than 1000 beach shacks (see Figure 9.13b), many built close to or on the beaches and in low-lying erosion and flood-prone areas. It will now be up to the government and taxpayers to maintain these unsightly ribbon developments, and to try and protect these properties as they become increasingly exposed to shoreline erosion and sea-level rise.

Western Australian coast

In contrast, the Western Australian government has been successively removing the many hundreds of beach and fishing shacks that dotted the coast north of Perth as far as Geraldton. These have been removed along with the associated myriad of 4WD tracks, and replaced by coastal reserves with well-planned and designed access points along the coast. Neighbouring coastal towns are being developed as nodes for the increasing coastal population.

Geographe Bay is a relatively sheltered, sandy embayment with a northerly drift of sand from Busselton north to Mandurah. In the early 1990s, this northerly drift was interrupted by the construction of a series of training walls and groynes associated with a canal estate called Port Geographe. Not only sand, but also piles of rotting seagrass built out 100 metres against the updrift wall, while the beach on the northern side eroded back 250 metres,

Figure 9.13 a) *Rock seawall along Adelaide's West Beach;* *b)* *Beach shacks at Lucky Bay beach are located in a low and hazardous position.* Photos: A.D. Short

Figure 9.14 *The Dawes Canal at Port Bouvard, which has a sand pumping system moving sand from the hooked southern capture groyne (right) to the northern side of the canal (left).*
Photo: A.D. Short

undermining a road and threatening houses. A combination of makeshift seawall and sand bypassing has been used with limited success; meanwhile, the rotting seagrass continues to pile up and waft across the development. Just 120 kilometres further north, the Port Bouvard development built a similar training wall, but also added a permanent sand bypassing system, the result being that no build-up occurs updrift and no erosion occurs downdrift (see Figure 9.14).

These few examples serve to illustrate that the coast is a dynamic and, at times, very inhospitable environment. When the coast is developed, it is essential that the natural processes and hazards are understood, including long-term rates of shoreline movement and change, and the extent of inland erosion and inundation. Planning must ensure that no inappropriate structures are placed in this hazardous zone, and that structures such as ports and airports are properly defended where they have to be. Where the longshore movement of sand is

interrupted, contingencies such as sand bypassing need to be in place, otherwise nature will realign the shores and place any downdrift development at risk.

The impact of climate change on the Australian coast

The Australian coast has experienced substantial changes of climate and sea level in the past, particularly in response to Quaternary glaciations and interglacials, as described in Chapter 1. It is also subject to considerable climate variability today, most notably in response to the changes that accompany the El Niño Southern Oscillation (ENSO) phenomenon. However, new concerns have emerged about the likely impacts of climate change as a result of global warming that threaten the coast in many ways (see Figure 9.15), including inundation of low-lying areas and accelerated coastal erosion. In many cases it will exacerbate human pressures on the coast.

Our understanding of climate changes in the Australian region is derived from trends over recent decades, and projections from global climate models. There is evidence for a general warming trend in air temperatures since 1910, indicating an average increase of 0.06 degrees Celsius per decade for mean maximum temperature and a greater increase of 0.12 degrees Celsius per decade for mean minimum temperature. Since the 1950s, there has been an increase in rainfall in the northwest associated with this warming, but a marked decrease in rainfall in southwestern Western Australia, and a decrease in eastern Australia.

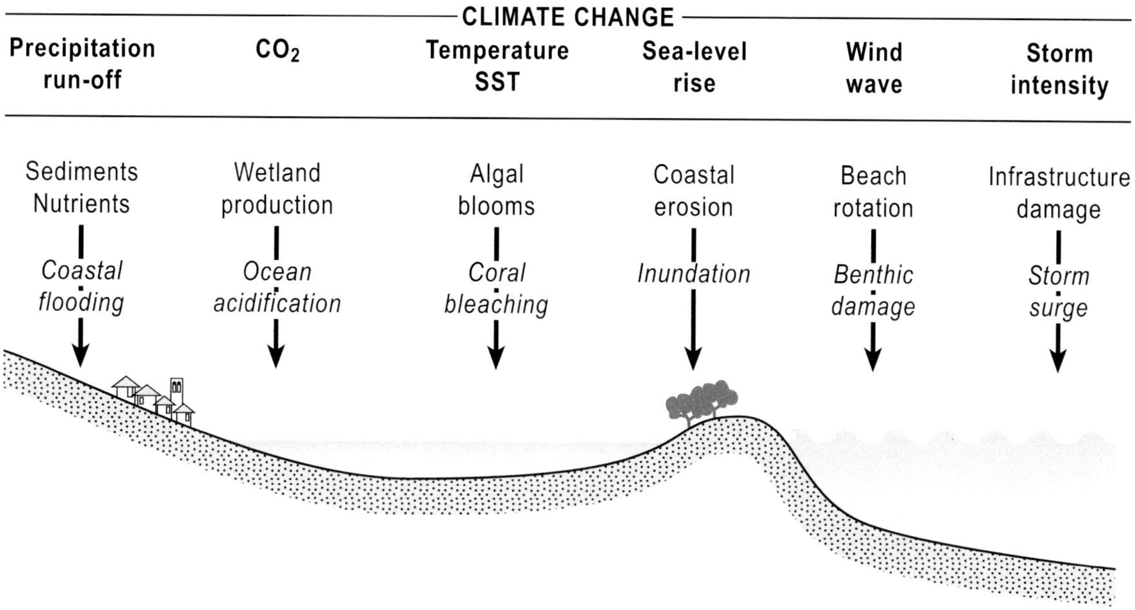

Figure 9.15 Climate change drivers and their impact on the coast, including chronic and acute drivers (italics)

Sea-level rise

Sea level is a particularly critical control on coastal behaviour. Tide gauges around the world indicate a global average rate of sea-level rise of 1.8 ± 0.3 millimetres per year for the period 1961 to 2003. There are only two tide gauge records of sufficient duration to undertake a long-term comparison in Australia, the gauge at Fremantle (commenced 1897) and that at Fort Denison in Sydney Harbour (commenced 1914). On the basis of these, sea level in Australia appears to have risen at 1.2 millimetres per year for the period 1920 to 2000. Based on satellite observations, the global rate of sea-level rise is calculated to have been 3.1 millimetres per year for the period 1993 to 2003. It is unclear whether this reflects acceleration in the rate of rise, or redistribution of water at a decadal scale as occurs in response to variability in the ENSO phenomenon.

Even if greenhouse gas concentrations can be stabilised during the 21st century, sea-level rise will continue because of the lag associated with the time it takes for the ocean to equilibrate with the climate. If warming eventually leads to the melting of most of the Greenland ice, and significant marginal melt of Antarctica, then a rise of several metres is possible, as is considered to have occurred during the Last Interglacial, 120 000 years ago.

Figure 9.16 shows the geographical variation in rates of sea-level rise as determined from tide gauges, but extrapolated using satellite altimetry, and the recorded rates of sea-level rise from a series of high-resolution Seaframe tide gauges that have been operating since the early 1990s. There is geographical variation between the records from different tide gauges and the disparity between the two indicates that there remains much uncertainty.

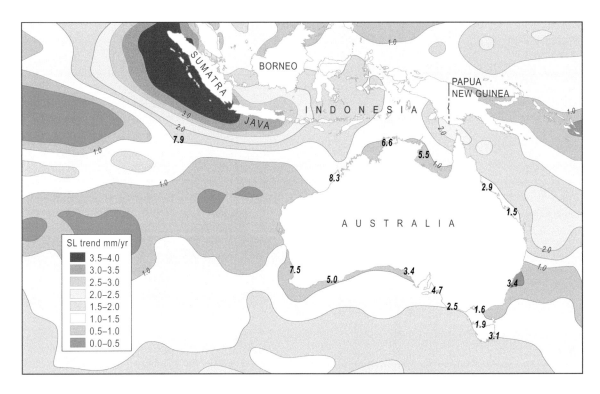

Figure 9.16 *Pattern of average rate of sea-level rise across the Australian region, based on more than 50 years of trend. The pattern of variability is determined from a short record of satellite altimetry, but the longer-term trends are derived from tide gauge records across the region. The rate of sea-level rise recorded by individual Seaframe tide gauges around Australia since their installation in the early 1990s is also shown.* Source: based on data from Church, Hunter, McInnes & White, 2006 and the National Tidal Centre

The highest rates of sea-level rise appear to be recorded in northern and western Australia; for example, Broome indicates a rise of more than 8 mm/year, while both Darwin and Perth have experienced rates of more than 6 mm/year.

Climate projections

Climate projections, based on global climate models, indicate warming over Australia of about 0.2 degrees Celsius per decade for the next two decades, and ongoing sea-level rise. By 2100, climate models indicate likely warming of between 2 and 4 degrees Celsius, and sea-level rise of between 18 and 59 centimetres, depending upon the rate of greenhouse gas emissions. The prime cause of the anticipated rise in sea level is thermal expansion of the ocean surface waters (~ top 200 metres of water) in response to warming, but the second most significant contributing factor is the melt of glaciers and mountain ice caps, with only a minor contribution from melt of the Greenland and Antarctic polar ice sheets. Low-lying islands and extensive coastal plains associated with estuaries and deltas, and all tidal flats are particularly at risk. These might be expected to experience increased levels of inundation and flooding, accelerated coastal erosion and saline intrusion into coastal waterways and water tables.

Rising sea level is likely to result in gradual shift of tidally zoned organisms upward and landward. On rocky, near-vertical shorelines, there may be little change except re-adjustment of intertidal organisms such as the tubeworm, *Galeolaria*. On low-gradient, soft sedimentary coastlines, the most probable long-term response of organisms tied to tide and sea levels is that they will experience a vertical and, in many cases, substantial horizontal shift in their habitat. In estuarine environments the response will depend upon the existing topography and sediment budgets (see Chapter 4). If the land is uninhabited, this is likely to result in landward migration of many intertidal and shallow subtidal communities such as seagrasses, mangroves and salt marshes as sea level rises. If there are built structures that prevent this migration, then these ecosystems are likely to be reduced by a process called coastal squeeze, whereby the organisms are unable to migrate landward.

The gradual rise of sea level will be imperceptible, but of particular concern is the effect that this will have on the extreme water levels. The level that a storm surge reaches is described in terms of its recurrence; that is, its average frequency over a long time. As the sea rises, particular extreme high-water levels will become more frequent. A storm surge level that floods the lowest parts of coastal cities, such as Cairns, and recurs on average once in 100 years, will become more frequent, both because sea level has risen and because it seems likely that storms will intensify. Parts of coastal towns in Queensland are likely to experience more frequent inundation.

Vulnerable coasts

Sea-level rise around Australia has been slight to date, but its acceleration is likely to exacerbate coastal erosion, particularly at sites where the natural responsiveness of the coast is reduced through human activities or intervention. Low-lying coastal plains, such as those associated with estuaries, are also threatened. The extensive wetlands bordering the many macrotidal estuaries along the coast of northern Australia appear to be especially vulnerable, such as those in the Alligator Rivers Region of Kakadu National Park (see Chapter 4). Several parts of the shoreline are already experiencing erosion, and the low-lying, sandy chenier ridges have been overwashed in several places, allowing incursion of seawater and incision by tidal creeks, the most extensive of which has been on the Mary River system (see Figure 9.17).

The plains in northern Australia are likely to see re-establishment of mangrove and salt marsh in many places as these intertidal habitats migrate landwards over high tide salt flats as a result of sea-level rise. In southern Australia there are fewer places where this will happen. Coastal squeeze is much more likely in southern Australian estuaries as obstacles such as roads, buildings or other structures prevent the landward migration of these intertidal ecosystems, and the tidal flats that support many of these ecosystems require thousands of years to accumulate.

Sensitive coastal ecosystems

Other aspects of climate change also threaten coastal systems, as indicated in Figure 9.15. Global warming is causing increases in sea-surface temperature. One of the most serious consequences is more frequent coral bleaching, discussed in Chapter 8. A widespread bleaching event occurred in 1998, causing serious damage to approximately 16 per cent of reefs around the world, with disastrous consequences in many parts of the Indian Ocean. On the Great Barrier Reef, 4 per cent of shallow water coral reefs experienced bleaching, and around 2 per cent died. Particularly extensive bleaching occurred again on the Great Barrier Reef in 2002; the widespread bleaching event affected 55 per cent of reefs, with mortality of around 5 per cent. Further bleaching also occurred in the summer of 2006.

Reefs and other calcareous organisms also appear threatened by ocean acidification, which is the lowering of pH that accompanies the increased absorption of carbon dioxide into the oceans. Ocean acidification results in decreased calcification in several major groups of organisms, including corals, and is likely to be a serious issue for reefs, with rates of calcification projected to fall by as much as 40–60 per cent under a doubling of pre-industrial levels of carbon dioxide, weakening corals and increasing their susceptibility to erosion.

Other marine and coastal ecosystems are susceptible to similar impacts. The greatest increases in sea-surface temperature in the southern hemisphere have already been observed associated with the East Australian Current (see Chapter 2). Increases of the order of 0.15 degrees Celsius per decade have been observed at 50 metres depth off Maria Island, eastern Tasmania. Ecological responses are likely to include a poleward movement of ecosystems; for example, a southward shift in the latitudinal limit to some species of seaweed has already been observed in New South Wales, and there has been reduction of kelp in northeastern Tasmania. Other changes, such as poleward extension of a species of sea urchin and an introduced species of crab, may also be related to increased sea-surface temperature. Further strengthening of the East Australian Current, including rises of 1–2 degrees Celsius by 2070, is projected by climate models and is likely to have several consequences, including poleward shift of pelagic fish such as Yellowfin Tuna, and other coastal species. Plankton are likely to respond quickly, with consequences such as more frequent appearance of harmful algal blooms in coastal waters. The time of important stages in the life cycle of organisms is also likely to change, with earlier migration of migratory species such as whales, or nesting of seabirds and waders.

Other aspects of climate change and their impacts

The shift in climate belts as a result of global warming is likely to mean changes in wave energy and direction. There has been a detectable increase in windiness and an increase in wave height of around 0.5 centimetres per year observed in the region. It is less clear how the incidence of storms will change in future. Although frequency may decrease slightly, increased sea-surface temperatures mean that the number of very intense storms seems likely to increase. The greater intensity of storms, particularly tropical and east-coast cyclones, means that higher waves and storm surges will lead to greater inundation. When combined with sea-level rise this is likely to result in considerably more frequent inundation of some low-lying areas, threatening the rapidly expanding coastal communities, particularly in the Cairns and southeast Queensland regions, while the New South Wales coast may be exposed to higher waves from the north and east. These will be further threatened by alterations to run-off in the catchment, which in the case of tropical storms mean that coastal flooding is exacerbated through the coincidence of flooding watercourses with elevated storm-surge water level at the coast.

Adaptation to climate change

The impacts of climate change are going to become more apparent on the coast; even if greenhouse gas emissions can be stabilised, there will continue to be some sea-level rise, and therefore it is important to consider adaptation measures. The adaptation options are:

1. to protect the shoreline by, for example, using coastal defences such as seawalls or repeated renourishment of beaches with sand
2. to accommodate, for example by raising the floor levels of buildings
3. to retreat, abandoning the shoreline to the rising waters and waves.

Figure 9.17 *A saline tidal creek cutting back into low-lying coastal plain of the Mary River that was covered by freshwater wetlands of paperbark trees that are now dead. The grey, muddy surface is a former channel, or palaeochannel, and the expanding tidal creek system is preferentially re-occupying these low-lying surfaces and re-establishing mangroves, a trend that may become more widespread as sea level rises in this area.* Photo: C.D. Woodroffe

Quite sophisticated plans for progressive response have been adopted in parts of Europe, including managed re-alignment where staged inundation of the sea into formerly reclaimed wetlands will be allowed to happen. In Australia, more than half the coast is composed of soft sediments (sand or mud), and hence at risk of erosion under future sea-level rise. However, much is remote and undeveloped, and would not justify protection.

In order to assess the level of risk, all spheres of government in Australia are now involved in a nationwide coastal vulnerability assessment, so we can be forewarned and prepared as the risks eventuate.

The energetic beaches on the wave-dominated coasts of southern Australia are subject to substantial storm cut when a large storm occurs. In the future it will not be the subtle and gradual upward creep of sea level that causes the greatest concern, but the impact of an extreme event, such as a major storm, where accompanying high waves and surges reach unprecedented elevations, accentuated by sea-level rise. The 1974 storms, for example, resulted in retreat of the shoreline by several tens of metres in southeastern Australia, and it took a decade or more for the shore to fully recover. This storm cut and recovery involves large sand volumes that will mask the gradual retreat that can be anticipated as a response to sea-level rise alone. Beach replenishment and the stabilisation of dunes with vegetation are among a series of potential adaptation strategies that may need to become more widely adopted in the Australian context (as in Europe and North America), in those situations where the natural behaviour of a shoreline becomes unacceptable. Particularly at risk are heavily capitalised shorelines, such as the Gold Coast, and the Adelaide metropolitan beaches, where such practices are already incorporated into the management of the shoreline, at great expense. Even on highly developed beaches such as Bondi and Surfers Paradise, the impact of sea-level rise could be retarded by decades using massive beach-sand nourishment.

Climate change will result in further stresses on coasts already impacted by the consequences of coastal development. There can be little doubt that much of tropical Australia will experience ferocious cyclones in the future, and there will be much debate about whether or not individual storms were more intense as a consequence of human-induced climate change, as has occurred following the tragic devastation caused in 2005 across New Orleans in the United States by Hurricane Katrina. Building codes and the lowest elevation at which buildings are permitted need to be increased appropriately in those places at risk. Ecosystems show considerable resilience in the face of natural processes, including extreme events, but adaptation involves maximising this resilience. The reduction of non-climate stresses on coasts can also serve to increase the resilience of coasts and reduce the impacts of climate change. For example, in the case of coral reefs, reducing the effects of overfishing and pollution, and other stresses such as sediment run-off, mean that some of the impacts associated with climate change may be less severe.

Conclusion: Prospects for the Australian coast

In this chapter we have traced the interaction of humans with the Australian coast. Australia was originally colonised by people whose origins remain beyond the Dreamtime. Early Europeans explored the coast and made contact with coastal Aborigines and islanders; however, their impact was transitory and their accounts were largely descriptive. Following the arrival of the First Fleet in 1788, coastal settlements expanded and more detailed coastal surveying, mapping and charting were undertaken, particularly to make shipping safer and to support a maritime tradition upon which Australia became dependent. Early contact between settlements relied on ships, and transport and communication was by sea. Although the 'outback' was gradually explored and settled to a limited degree, the coast has remained the focus of Australian trade and life. Most of the population lives along the coast; the coast is where many Australians undertake their recreational activities and it is the focus of most domestic tourism and the majority of international tourism.

What future does the Australian coast face? The pressures are twofold – human and natural, the latter also linked to climate change. The former can be managed and controlled to the degree that society demands. With wider adoption of the sea-change lifestyle, there will be growing pressure for increased coastal development and at higher density.

The second challenge is natural, and particularly the challenge of climate change and its associated impacts on the coast, which is predicted to include sea-level rise, increased tropical cyclone intensity and changing wave, wind and rainfall patterns. Sea-level rise will result in gradual retreat and erosion of the 50 per cent of the coast composed of sandy beaches. There will be a demand to preserve and protect the more popular 400–500 beaches, at most. This can take the form of beach sand nourishment, particularly along the southeast coast, where huge sand reserves lie just off the coast, with seawalls and other structures as a last resort. The backing dune systems will be potentially reactivated by shoreline erosion; however, these can be managed with dune fencing and planting.

The 40 per cent of the coast that is rocky and generally of high relief will be left to cope without interference, other than managing human use of these locations, which will become increasingly hazardous as sea level rises. The deltas, floodplains, estuaries and associated wetlands will bear the largest impact, as even slight rises in sea level will produce considerable physical and ecological impacts. The 5000 kilometres of open-coast tidal flats plus the even greater extent of low-lying estuarine salt marshes, mangroves and seagrass meadows will be gradually inundated, which will result in a landward shift of these systems where possible. However, most of the tidal flats have taken thousands of years to evolve, and so in low-energy and estuarine environments there will be a reversion to former deeper, more open-water conditions, and in many areas a shrinking of the intertidal salt marshes and mangroves. Likewise the coral reef systems that fringe northern Australia will have to catch up with rising sea level or be left behind. As climate changes, the Australian, and the world's, coasts face an interesting and uncertain future. Fortuitously, in Australia, which has a large, mostly undeveloped coast, nature will accommodate and respond to the changes as it has in the past. We will need to soften the impact with beach nourishment along our popular beaches, build seawalls and barrages to protect important infrastructure and property, and elevate our existing ports and coastal airports out of the reach of the sea.

The present challenge is how to achieve these coastal planning and management goals while an increasing number of Australians want to live and play at the coast, and national and international tourists want to be accommodated as close to the shore as possible. To effectively manage this demand, coastal managers require detailed, high-resolution environmental information, in order to assess the vulnerability of the coastal zone and its ecosystems to both human pressure and natural hazards. This information should feed into coastal policies that are effectively managed and enforced, ensuring that the coast and its habitats are protected and that development is excluded from coastal hazard zones and sensitive, valuable coastal environments and ecosystems. Most importantly, coastal development should be contained and constrained, focusing new development in existing coastal settlements, so as to avoid ribbon development. At the same time, large sections of coast and adjoining coastal waters need to be preserved in coastal national parks and marine protected areas. The policies also need to be cognisant of the potential impacts of climate change, so that coastal habitats have room to move, and human occupation is not put at risk.

The above challenges are well within the reach of Australian society. We already have an Australia-wide coastal policy, as well as state/territory coastal agencies, policies and management plans. The Australian coast is in good hands and Australians are widely regarded as world leaders in coastal science and coastal management. However, we do have a massive coastline, one that is highly variable and in many areas extremely valuable, and that is under growing pressure from society, recurring natural hazards, and increased future threats. Its effective management requires a thorough understanding of how its physical and biological systems operate, based on detailed scientific studies. The results of these can be used to underpin effective coastal management plans. The challenge ahead is for Australia to continue scientific investigation of our coastal systems, to effectively review and fine-tune the coastal policy and management plans, and to ensure the policies are enforced so that future generations may enjoy the benefits of the coast. For the first time in history we have the opportunity and ability to leave the coastal zone in a better condition than we found it.

FURTHER READING

Print materials

Birch, W.D. (ed.), 2003, *Geology of Victoria*. Geological Society of Australia, Victoria Division, Special Publication 23, 842 pp.

Bird, E.C.F., 1993, *The Coast of Victoria*. Melbourne University Press, Melbourne, 324 pp.

Brearley, A., 2005, *Swanland – estuaries and coastal lagoons of south-western Australia*. University of Western Australia Press, Perth, 550 pp.

Burrett, C.F. and Martin, E.L. (eds), 1989, *Geology and Mineral Resources of Tasmania*. Geological Survey of Tasmania, Special Publication 15, 574 pp.

Butler, A.J. and Jernakoff, P., 1999, *Seagrass in Australia: strategic review and development of an R&D plan*. CSIRO Publishing, 210 pp.

Chappell, J. and Woodroffe C.D., 1994. 'Macrotidal estuaries'. In R.G. Carter and C.D. Woodroffe (eds), *Coastal Evolution: Late Quaternary shoreline morphodynamics*. Cambridge University Press, pp. 187–218.

Church, J.A., Hunter, J.R., McInnes, K.L. and White, N.J., 2006. 'Sea-level rise around the Australian coastline and the changing frequency of extreme sea-level events.' *Australian Meteorological Magazine*, 55, pp. 253–60.

Davies, J.L., 1980, *Geographical Variation in Coastal Development*. 2nd edn, Longman, London, 212 pp.

Davis, R. and FitzGerald, D., 2003, *Beaches and Coasts*. Malden Maryland, Blackwell, 419 pp.

Day, R.W., Whitaker, W.G., Murray, C.G., Wilson, I.M. and Grimes, K.G., 1983, *Geology of Queensland*. Geological Survey of Queensland, Brisbane, Publication 383, 194 pp.

Drexel, J. F., Preiss, W. V. and Parker, A. J., 1995, *The Geology of South Australia. Volume 1: The Precambrians*. South Australian Geological Survey, Bulletin 54, 242 pp.

Drexel, J.F. and Preiss, W.V., 1995, *The Geology of South Australia. Volume 2: The Phaneroizoic*. South Australian Geological Survey, Bulletin 54, 347 pp.

Duke, N.C., 2006, *Australia's Mangroves*. University of Queensland, Brisbane, 200pp.

Edgar, G.J., 2000 *Australian Marine Life: The plants and animals of temperate waters*. Reed New Holland, Sydney, 544 pp.

Furnas, M., 2003, *Catchments and Corals: Terrestrial runoff to the Great Barrier Reef*. Australian Institute of Marine Science, Townsville. 334 pp.

Geological Survey of Western Australia, 1990, *Geology of Western Australia*. Western Australia Geological Survey Memoir 3, 827 pp.

Groves, R.H., 1994, *Australian Vegetation*. Cambridge University Press, 562 pp.

Harris, P.T., Heap, A.D., Bryce, A.D., Porter-Smith, R., Ryan, D.A. and Heggie, D.T., 2002, 'Classification of Australian clastic coastal depositional environments based upon a quantitative analysis of wave, tidal, and river power.' *Journal of Sedimentary Research*, 72, pp. 858–70.

Harvey, N. and Caton, B., 2002, *Coastal Management in Australia*. Oxford University Press, Port Melbourne, 342 pp.

Heap, A.D., Bryce, S.M., Ryan, D.A., Radke, L.C., Smith, C.S., Harris, P. and Heggie, D.T., 2001. *Australian Estuaries and Coastal Waterways – A geoscience perspective for improved and integrated resource management*. Geoscience Australia Record 2001/07, http://dbforms.ga.gov.au/pls/www/npm.Ozcoast.show_mm?pBlobno=12885

Henderson, R.A. and Stephenson, P.J. (eds), 1980, *The Geology and Geophysics of Northeastern Australia*. Geological Society Australia, Queensland Division, Brisbane, 468 pp.

Further reading

Hesp, P.A., 1999, 'The beach backshore and beyond.' In A.D. Short (ed.) *Handbook of Beach and Shoreface Morphodynamics*. Wiley, Chichester, pp. 145–69.

Hopley, D., Smithers, S.G. and Parnell, K.E., 2007, *The Geomorphology of the Great Barrier Reef: Development, diversity and change*. Cambridge University Press, Cambridge, 532 pp.

House of Representatives Standing Committee on Environment, Recreation and the Arts. 1991, *The Injured Coastline: Protection of the coastal environment*. Canberra: Australian Government Publishing Service.

Johnson, D., 2004, *The Geology of Australia*. Cambridge University Press, Melbourne, 276 pp.

Johnson, J.E. and Marshall, P.A. (eds), 2007, *Climate Change and the Great Barrier Reef*. Great Barrier Reef Marine Park Authority and Australian Greenhouse Office, Australia, 818 pp.

Kay, R. and Alder, J., 2005, *Coastal Planning and Management*. Taylor & Francis, London, 380 pp.

Laughlin, G., 1997, *The User's Guide to the Australian Coast*. New Holland, Sydney, 213 pp.

Morwood, M.J., 2002, *Visions from the Past: The archaeology of Australian Aboriginal art*. Allen & Unwin, 347 pp.

Parry, M.L., Canziani, O.F., Palutikof, J.P., van der Linden, P.J. and Hanson, C.E. (eds), 2007, *Climate Change 2007: Impacts, adaptation and vulnerability*. Contribution of Working Group II to the Fourth Assessment Report of the Intergovernmental Panel on Climate Change, Cambridge University Press, Cambridge, UK.

Pearson, M., 2005, *Great Southern Land: The maritime exploration of Terra Australis*. The Australian Government, Department of the Environment and Heritage, 155 pp.

Resource Assessment Commission. (1993). *Coastal Zone Inquiry – Final report*. Canberra: Australian Government Publishing Service, available from www.environment.gov.au/coasts/publications/rac/preface.html, accessed 14 September 2008.

Roy, P.S., 1984, 'New South Wales estuaries: Their origin and evolution.' In B.G. Thom, *Coastal Geomorphology in Australia*. Sydney, Academic Press, pp. 99–121.

Roy, P.S., Cowell, P.J., Ferland, M.A. and Thom, B.G., 1994, 'Wave-dominated coasts.' In R.W.G. Carter and C.D. Woodroffe (eds), *Coastal Evolution: Late Quaternary shoreline morphodynamics*. Cambridge University Press, pp. 121–86.

Roy, P.S., Williams, R., Jones A.R, Yassini, I., Gibbs, P.J., Coates, B., West, R.J., Scanes, P.R., Hudson. J.P. and Nichol, S., 2001, 'Structure and function of south-east Australian estuaries.' *Estuarine Coastal and Shelf Science*, 53, pp. 351–84.

Ryan, D.A., Heap, A.D., Radke, L.C. and Heggie, D.T., 2003. *Conceptual Models of Australia's Estuaries and Waterways. Applications for coastal resource management*. Geoscience Australia Record 2003/9.

Scheibner, E. and Basden, H., (eds), 1995, *Geology of New South Wales. Volume 2: Geological evolution*. Geological Survey of New South Wales Memoir 13. 666 pp.

Scheibner, E. and Basden, H. (eds), 1995, *Geology of New South Wales. Volume 1: Structural framework*. Geological Survey of New South Wales Memoir 13. 295 pp.

Short, A.D., 2007, *Beaches of the New South Wales Coast*. 2nd edn, Sydney University Press, Sydney, 398 pp.

Short, A.D., 2006a, *Beaches of the Tasmanian Coast and Islands*. Sydney University Press, Sydney, 353 pp.

Short, A.D., 2006b, *Beaches of Northern Australia: The Kimberley, Northern Territory and Cape York*. Sydney University Press, Sydney, 463 pp.

Short, A.D., 2005, *Beaches of the Western Australian Coast: Eucla to Roebuck Bay*. Sydney University Press, Sydney, 433 pp.

Short, A.D., 2001, *Beaches of the Southern Australian Coast and Kangaroo Island*. Sydney University Press, Sydney, 346 pp.

Short, A.D., 2000, *Beaches of the Queensland Coast: Cooktown to Coolangatta*. Sydney University Press, Sydney, 360 pp.

Short, A.D. (ed.), 1999, *Handbook of Beach and Shoreface Morphodynamics*. John Wiley and Sons, Chichester, 379 pp.

Short, A.D., 1996, *Beaches of the Victorian Coast and Port Phillip Bay*. Sydney University Press, Sydney, 298 pp.

Stephenson, W.J. and Kirk, R.M., 2005, 'Shore platforms'. In M. Schwartz (ed.), *The Encyclopedia of Coastal Science*. Kluwer Academic Publishers, pp 873–5.

Stephenson, W.J. and Thornton, L.E., 2005, 'Australian rock coasts: Review and Prospects.' *Australian Geographer, 36*, pp. 95–115.

Thom, B.G. and Harvey, N., 2000, 'Triggers for late twentieth century reform of Australian coastal management.' *Australian Geographical Studies, 38*, pp. 275–90.

Turner, L., Tracey D., Tilden, J. and Dennison, W.C., 2004, *Where River Meets Sea: Exploring Australia's estuaries*. Cooperative Research Centre for Coastal Zone, Estuary and Waterway Management, CSIRO Publishing.

Underwood, A.J. and Chapman, M.G. (eds), 1995, *Coastal Marine Ecology of Temperate Australia*. University of New South Wales Press, Sydney, 341 pp.

Veron, J.E.N., 2008, *A Reef in Time: The Great Barrier Reef from beginning to end*. The Belknap Press of Harvard University Press, Cambridge Massachusetts, 289 pp.

Veron, J.E.N., 1995, *Corals in Space and Time: The biogeography and evolution of the scleractinia*. Sydney, University of New South Wales Press.

Woodroffe, C.D., 2003, *Coasts: Form, process and evolution*. Cambridge University Press, Cambridge, 623 pp.

Woodroffe, C.D., Chappell, J., Thom, B.G. and Wallensky E., 1989. 'Depositional model of a macrotidal estuary and floodplain, South Alligator River, Northern Australia.' *Sedimentology, 36*, pp. 737–56.

Woodroffe, C.D., Mulrennan M.E. and Chappell J., 1993, 'Estuarine infill and coastal progradation, southern van Diemen Gulf, northern Australia.' *Sedimentary Geology, 83*, pp. 257–75.

Zann, L.P. (ed.) 1995, *Our Sea, Our Future: Major findings of the Stae of the Marine Environment Report for Australia*. Canberra: Department of the Environment, Sport and Territories and Townsville, Qld: Great Barrier Reef Marine Park Authority, available at www.environment.gov.au/coasts/publications/somer/index.html, accessed 14 September 2008.

Websites

Australian Beach Books

www.sup.usyd.edu.au/marine

This University of Sydney Press site has information on how to order books that describe the beaches of every state and cover every beach in Australia.

Australian Beach Safety and Management Program

www.slsa.com.au/default.aspx?s=beachmgmtandsafetyprog

An overview of the Australian Beach Safety and Management Program, which contains data on all Australian beaches.

Australian Bureau of Meteorology

www.bom.gov.au and www.bom.gov.au/oceanography

The Australian Bureau of Meteorology provides a range of weather and climate services, including details of past tropical cyclones, as well as the National Tidal Centre, which issues tidal predictions and real-time tidal observations from the Seaframe high-resolution tide gauges. It also analyses climate change, including Australian climate change and variability. The Bureau's oceanography section provides marine and ocean weather forecasts, as well as tide prediction and tsunami warnings.

The Australian Coastal Safety Resource

www.AussieBeach.com.au

An online resource that provides information on safety issues, coastal amenities and infrastructure, and tourist destinations and services. The current focus is on southwestern Western Australia. A great resource for beachgoers when fishing, surfing, swimming, snorkelling or just visiting the coast.

Australian Institute of Marine Science

www.aims.gov.au

Australia's tropical marine research agency provides details of research and resources related to northern reefs and publications.

Buoy Weather

www.buoyweather.com

This site provides wind and wave forecasting for the world.

Coastalwatch

www.coastalwatch.com

This site operates a series of surf cams around the nation and overseas, an automated, real-time coastal observation system and data service for coastal monitoring.

Department of Climate Change

www.climatechange.gov.au/impacts/coasts.html

This site has a section dedicated to the impact of climate change on the coast, including information on coastal research and reports.

Department of Environment, Water, Heritage and the Arts

www.environment.gov.au/coasts

This site has a section on coasts and oceans, including a number of publications, a section on marine protected areas, and the Australian Coastal Atlas, which incorporates a variety of interactive mapping tools.

Geoscience Australia

www.ga.gov.au

Geoscience Australia provides data, technical information, advice and research for maritime boundary definition, regional marine planning and environmental management.

The Great Barrier Reef Marine Park Authority

www.gbrmpa.gov.au

The website of the management agency that oversees the Great Barrier Reef provides a gateway for information about the Great Barrier Reef and World Heritage Area, including plans of management and zoning.

Manly Hydraulics laboratory

www.mhl.nsw.gov.au

This site provides real-time data from the New South Wales waverider buoys and tide gauges.

New South Wales beaches

www.visitnsw.com/Beaches_p640.aspx

The official tourist site for many of the state's beaches.

OzCoasts

www.ozcoasts.org.au

OzCoasts provides comprehensive online information about Australia's coast, including its estuaries (formerly OzEstuaries), beaches and coastal waterways.

Queensland Environment Protection Agency

www.epa.qld.gov.au/environmental_management/coast_and_oceans

The Queensland Government's site for information about the state's coastal environment, including beaches and dunes, marine habitats, the Great Barrier Reef, coastal management, and wave and storm tides.

South Australia Atlas

www.atlas.sa.gov.au

An online atlas of South Australia, including all its coastal regions.

Surf Life Saving Australia

www.slsa.asn.au

Surf Life Saving Australia provides comprehensive information on all aspects of surf lifesaving, as well as links to all state/territory centres and surf lifesaving clubs. It also has a 'Find a beach or club' section, which locates and describes every surf lifesaving club and patrolled beach in Australia.

Victorian Coastal Council

www.vcc.vic.gov.au

This site provides information about the strategic planning and management of the Victorian coast.

INDEX

Printed in the United States
By Bookmasters